The Balance of Nature

In this exciting introduction to the world around us —which does for nature what Rachel Carson did for the sea in The Sea Around Us—a noted authority explains the pattern of life of plants and animals and clearly shows how all living things are related to each other and to their environment.

You will learn how plants build food from water, air, and sun; how animals use plants for food; and how animals, in turn, breathe out the carbon contained in plants to keep the life cycle constantly moving. With magnificent clarity, the author explains the unending progression of life caused by the survival of the fittest, the evolution of new species that can more readily cope with environmental changes, and the dynamic transformation wrought by man on nature.

This fascinating book tells the vital and interesting story of how the natural world operates and how man can maintain this delicate balance of nature by using modern conservation methods.

"Fascinating reading . . . It is a volume that all thinking men should read—and it is a pleasure to be able to say that they will enjoy it"
—*CHICAGO TRIBUNE*

the web

With an Introduction by FAIRFIELD OSBORN

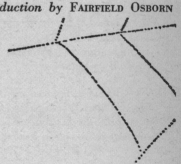

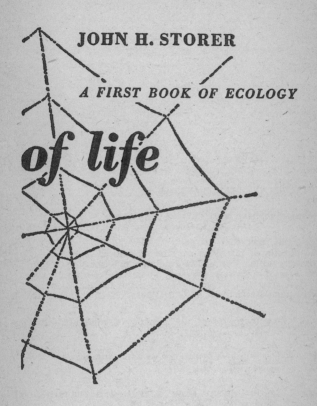

JOHN H. STORER

A FIRST BOOK OF ECOLOGY

of life

A MENTOR BOOK

NEW AMERICAN LIBRARY

TIMES MIRROR

NEW YORK AND SCARBOROUGH, ONTARIO

THE NEW ENGLISH LIBRARY LIMITED, LONDON

MENTOR TRADEMARK REG. U.S. PAT. OFF. AND FOREIGN COUNTRIES
REGISTERED TRADEMARK—MARCA REGISTRADA
HECHO EN CHICAGO, U.S.A.

SIGNET, SIGNET CLASSICS, MENTOR, PLUME AND MERIDIAN BOOKS
are published *in the United States* by
The New American Library, Inc.,
1301 Avenue of the Americas, New York, New York 10019,
in Canada by The New American Library of Canada Limited,
81 Mack Avenue, Scarborough, 704, Ontario,
in the United Kingdom by The New English Library Limited,
Barnard's Inn, Holborn, London, E.C. 1, England.

15 16 17 18 19 20 21 22 23

introduction

There are some truths, even fundamental ones, that are apt to elude us. The most basic truth regarding our Earth-home is that all living things, in some manner, are related to each other. This fact, while mainly important as a physical principle, carries implications even of a spiritual nature.

This book is important primarily because it is a study of the interrelationships of living things. It deals with a field of knowledge known as ecology. Sound conservation practices are based upon this knowledge and consequently this book is essential reading for everyone who is interested in conservation.

The Web of Life has been written by a man uniquely qualified for the task. Long years of first-hand observation in the field impressed upon John Storer the truth that no living thing, whether plant or animal, is independent. During earlier years, when he was accomplishing brilliant work as a wildlife observer and photographer, he became increasingly aware that environment was the dominant force and that all living species were subject to its conditions, or even the conditions of seemingly unimportant elements within the natural environment. How self-evident a fact, and yet how difficult for any of us, living under modern conditions, fully to understand. The aim of this book is to make that compelling truth clearer. The youngster, captive on the sidewalks of our big cities, the farmer struggling in a dust bowl, the sullen river that once ran silver, the desolate tangle of second growth, even the last condor on a California mountaintop—all have a tenuous relationship to life on this earth as a whole. Man does not stand alone.

Quite apart from the author's technical knowledge of the subject, his writing contains the qualities that are only derived from patient field observations of nature at work. No one could write such a book as this without a truly observant eye as well as a truly understanding heart. Without these qualities, *The Web of Life* would prove just another useful treatise on ecology. The present reader will carry away with him an enduring conviction that the basic principles and laws governing the entire living community must be generally understood if we human beings are to be successful in maintaining the productivity of the earth. Nature will not accept ignorance of her laws as an alibi.

Although the balance of nature is a complex business, the story told here is in simple language and presented with clarity. While this book is not written primarily for specialists, it is valuable for all students of agriculture, and even for students who are interested in the social sciences. One of our great ecologists, the late Aldo Leopold, became eminent in his field not only because he was an accomplished scientist but because he was a philosopher as well. He used to say that unless one approached conservation with an ethical as well as an economic perspective, the problem had not even been adequately defined. In this book, in referring to man's conquest of the natural world, the author writes:

"Under the domination of his intellect, the world's life, and the environment of that life, seem to have reached a crossroads, and the choice of direction is for the future to decide. . . . The great and deciding test, however, still remains—whether man can coordinate knowledge into understanding and build within his heart the incentives and the wisdom to use these new-found powers wisely, and with responsibility, for the common good."

It is a deep pleasure to have been asked to write this brief introduction, which carries the hope that this book may receive wide attention. It is a significant contribution to an understanding of what Conservation means to the future of our country and indeed to people the world over.

FAIRFIELD OSBORN

The Conservation Foundation
New York

acknowledgments

The first of my acknowledgments is to the Conservation Foundation, and to Mr. Fairfield Osborn and Mr. George Brewer, whose constructive interest in conservation made possible invaluable preparatory experience. I am especially indebted to Mr. Charles W. Mattison and to many other members of the U. S. Forest Service, the Soil Conservation Service, the National Park Service, the Fish and Wildlife Service, and the U. S. Weather Bureau.

These acknowledgments would not be complete without a tribute to the late Joseph S. Seabury, artist and naturalist, whose quick eye and keen understanding helped to open the way for my interest in the subject.

Thanks are also due to Professor H. I. Featherly, Mrs. Luis J. Franke, Dr. Edward H. Graham, the Rev. John La Farge, S.J., Mr. Roger Tory Peterson, Mr. Richard H. Pough, and Dr. Paul Sears for their interest in reading through the manuscript at various stages, and for many helpful suggestions.

I also wish to thank Messrs. Charles Scribner's Sons for permission to reproduce the magnificent passage from Alan Paton's *Cry, the Beloved Country*, which serves as a fitting closing for my book.

And, finally, I wish to dedicate the book to my wife Elizabeth and my daughter Ethel.

J. H. S.

contents

foreword: why this book?

This book is written for everyone interested in the wise use of our soil and water, our forests and wildlife, and for everyone who would learn what has largely been forgotten in our machine age—how all living things fit together into a single pattern.

The book undertakes to separate certain significant threads from a mass of facts and forces that go into the making of life's physical environment, and to crystallize them into a clear and comprehensive picture—a picture intended to show a natural community in action; how it grows, builds its own environment, and changes it until the plants and animals that started it can no longer live there and must give way to others better adapted to live in the newly created surroundings. We see the orderly progression of such changes until at last a condition of balance, or "climax," is reached.

We see the roots, insects, soil bacteria, and worms building fertility into the soil, and the shrews and moles that hunt these smaller creatures, "plowing" and aerating the soil, turning it into an absorbent sponge. . . .

A snowflake falls on the branch of a spruce tree on a mountainside. Part of it evaporates; part flows down to the roots and helps to support the forest that builds and protects the soil. The rest seeps down the hillside through a prepared and porous soil, forming with millions of its fellows a slowly moving reservoir that maintains the flow of brooks and rivers throughout the year.

Human communities in the valleys compete for this water, some of them even tunneling through mountains to divert its course over a radius of a thousand miles, until at length the growth of communities is stopped by the limits of the available water supply. The best use of that water and of the watershed that controls it has become a matter of life and death to those communities. The use and pollution, the cleansing and re-use of the water, in an effort to extend its usefulness, become matters of vital moment, demanding close cooperation between the people dependent on it. . . .

Or consider the case of a great spruce forest which, through a catastrophic windstorm, lost the protection of its woodpeckers. With this protection gone, the forest is killed by bark beetles, normally rather innocuous members of the community. The tinder-dry, dead forest then becomes a prey to fire, which destroys it and, along with it, the watershed. Resulting floods and droughts damage soil, ruin farms, and inundate cities a thousand miles away. . . .

Human intelligence, which has learned how to build, to outgrow, and to destroy its environment, raises additional questions. Take for example, a case in southern Arizona, where a small green pasture of rich native grasses is surrounded by an endless desert of sand and cactus. This whole desert was once an empire of grass as lush as the green pasture; but it was over-grazed and destroyed by cattlemen, who robbed and ruined land that might have supported generations of their descendants.

The small green pasture was properly grazed. It was protected by a strand of wire and, with the addition of improved grasses, is today more productive than ever. That single strand of wire had fenced out the cactus, the sand, and the eroding gullies, because it controlled the surplus cattle that prepared the way for them. It was put there by a man with understanding, who knew how to use his land and to maintain its productivity.

In conclusion, we may well ask whether man will develop understanding before he destroys himself by destroying his environment. The way would seem to lie in the building of an informed public interest in certain fundamentals.

I believe this book to be the first of its kind in this field, and of immediate interest to every intelligent citizen.

Winter Park, Florida   JOHN H. STORER

NOTE TO THE READER:

The illustrations in the middle of this book complement the text; they, and their captions, may be studied as a separate section, either before or after reading the text.

1. *the stream of life*

We once stood on a hillside in Wyoming and looked across the broad expanse of a lake backed by a rugged range of mountains. Their peaks stood silhouetted black against the late afternoon sun, and under its rays the lake glowed like a sea of liquid fire.

Suddenly, out from the shadow of the distant peaks a tiny black speck moved into the light on the lake's surface. As it moved toward us a trail streamed out behind it, and at last the shape grew into a speeding motor boat with the long lines of its wake sweeping out on either side, dark ribbons on the glowing water.

Finally, as it neared our side of the lake, the motor boat circled twice in wide sweeps and then disappeared behind a headland. On the lake behind it the lines of these circling wakes came together in interweaving patterns. In some places they collided to efface each other. In others they reinforced one another, finally rolling off in long curves to the distant shores.

These rolling waves were set in motion by the energy of the moving boat, that displaced the water in its passing. Each wave carried within itself a portion of that energy, finally expending it as it rolled a pebble on the shore, or rocked the roots of a floating derelict tree, or perhaps washed out the final handful of dirt from the undermined roots of a tree that would topple into the water to join the other derelict. Thus, the energy of the motor boat had its impact on the land far beyond the vision of the man who set it moving. Just so, the forces set in motion by every act of man or bird, animal, insect, or bacterium move out to affect the lives of many other creatures.

The principles which govern all these interrelationships are called the principles of ecology—the science which deals with the mutual relations between living organisms and their environment. The subject of ecology is so vast and complex that no human mind has ever fathomed all its secrets. Many of them can probably never be unraveled, but the basic principles of ecology are known, and on the functioning of these known principles depends the future of all human lives.

Life is such a very personal thing, wrapped up within the being of every living creature, that it is sometimes hard to realize how intimately each life is connected with a great many other lives.

Life is a flowing stream, forever passing away and as constantly being renewed. The energy that brings us life is supplied from many different sources, most of them beyond our vision of experience.

The bread that comes from the grocery shelf brings minerals and vitamins prepared by wheat plants in the soil of Kansas. Vegetables bring their special contributions from the farms of Texas or California. Meat was built by cattle, out of grass that drew its special qualities from the soil of the western plains. It required five thousand pounds of water to produce the grass that went into the making of each single pound of meat. To store this water and supply it to the grass as needed through summer droughts, the soil had first to be prepared by many generations of earlier plants and animals.

Bread, vegetables, and meat are merely vehicles for transferring to us the special properties of soil, air, and sunlight gathered and organized by the plants. With some minor variations all these living things are composed of the same elements, and in roughly the same proportions. But each differs in its methods of obtaining and using them. Plants and animals vary greatly in the amounts of light, heat, and moisture they must have for their development and in the kinds of soil on which they thrive best. Each is adapted to its own special environment.

In our modern civilization we find the stream of our own lives flowing from many different sources, each adapted to make its own particular contribution. Why should the citizen of New York or Pennsylvania want to draw on the soils of Texas or Kansas, Wyoming, or other far-off places for the minerals and vitamins that go into the building of his life? The answer is that these soils are more richly supplied with the necessary elements. Some of them did not originally exist in these spots but were transported and placed there by a number of different agencies.

The story of their coming, to make possible our own stream of life, lies deep in the past history of the earth.

2. *what the rocks contribute*

There is very much that we do not know about the earth's early history, and our knowledge of its interior today is far from definite. But there is good evidence that the surface of the globe was at one time a mass of molten material. Within this molten mass the heavier elements such as iron and nickel sank deep below the surface, the lighter elements floating over them.

Slowly this liquid mass cooled, and as it did the elements that composed it tended to draw together, each to its own kind, to form crystals—silica forming quartz crystals, for example—or they gathered into groups to form more complex minerals, such as mica or hornblende.

Over most of the earth's original crust, this grouping of crystals took the form of granite. Granite ordinarily carries few of the elements needed for life—potassium being one of the few—and is a poor environment for a plant. But this thin crust of granite lay over an uneasy foundation of molten rock. As the hot mass cooled and contracted, great ridges were forced up to form mountains, and in some places huge cracks were opened through which the molten rock from below flowed up to spread in sheets of lava over the lighter outer layers of rock. Smaller cracks were filled with molten material to form veins in the rock. In some cases, vaporized metals from below were cooled and solidified in these veins, and boiling water, carrying elements in solution, welled up from the inner earth, to deposit them in the cracks.

Thus, some parts of the earth's surface became rich with concentrations of rare minerals that were gifts from the deep interior of the earth. In many parts of the world, where hot springs may be seen boiling to the surface, this process is still going on.

These concentrations of minerals, if left undisturbed, would not have supplied a very wide area with the essentials of life. But the forces of nature are constantly at work to break up the rocks and spread them over the earth's surface. Air, water, and sunlight all do their part. Air supplies carbon dioxide. Falling rain absorbs this gas from the air to form carbonic acid. Through the years this acid slowly dissolves the more soluble minerals from the rocks, leaving crevices where water may enter. The sun warms the exposed faces of the rock, causing them to expand and crack, making room for more water. When water freezes it expands, so in cold weather the cracks are widened, sometimes breaking up huge blocks of rock. Thus the rock begins to disintegrate as soon as it is exposed to the weather until, at last, it is reduced to the small particles that go into the making of soil. Much of this soil material lies in place, where it is born. But a good share is scattered far and wide by other forces of nature.

Wind and flowing water sweep rock fragments from the ridges, drive them against other exposed rocks, grind them to smaller pieces, and sometimes carry them for many miles to drop them at last as sediment in distant valleys or in lake or ocean beds.

Now the water may change from destroyer to builder. As it carries the dissolved minerals in solution, it may deposit some of them among these sediments, thus cementing them into rock again.

Microscopic living organisms absorb minerals from

these liquid solutions, too, and turn them to solids within their substance. Through the ages their remains have settled on the bottom to form vast deposits of limestone and phosphates.

Later on, through movements of the earth's crust, some of these deposits have been lifted high above the water. Many of our western mountains are capped by rocks holding fossils of shells that were made of lime taken from the water of an ancient sea. These rocks are now attacked in their turn by sun and frost and rain, finally to be scattered over the earth by wind and water. In past ages enormous quantities of rock have been ground to dust by the ice of moving glaciers and carried many miles before being added to the material of the soil.

It is estimated that there is scarcely a square mile of the earth's surface that does not contain some ingredient from every other mile, because of the action of wind and water. But, despite all this stirring and mixing, the soil usually draws most of its mineral content from the underlying parent rock that produced it. Where this rock is rich in important minerals, as in the case of limestone, the soil can usually provide a rich environment for life. But over rocks like granite or quartzite, which offer little mineral food, the life above is apt to find a much less nourishing environment.

3. *how the land gets its water*

The upheavals and varied contours on the earth's surface spread out their influence to affect our lives in many ways. They are among the important factors in nature's water transportation system that makes life possible on the land.

Among its many useful qualities, water has one that is especially valuable to life. When it is warmed above a certain temperature, it turns to vapor and can be carried by the wind. Warm air will absorb a lot of water vapor. When the air is cooled, some of this vapor turns back to fine drops of water in the form of clouds and under certain conditions precipitates as rain.

The temperature of the air is constantly changing from day to night and with the seasons. While the sun's rays pass through the air without raising its temperature, they warm the surface of the earth and the oceans very noticeably. Air coming in contact with these heated surfaces is then warmed.

The result is that air near the earth's surface is usually much warmer than the air high overhead. This added warmth has two dramatic effects on air. It not only gives it greater power to absorb moisture, but under its influence air expands and becomes lighter. Light air rises through heavy, cooler air above, while cool air sinks to replace it at the surface. So there is a steady circulation of air, rising from the warmer parts of the earth, only to be cooled in the upper atmosphere and return to earth again.

As warm air rises from the ocean surface, it is laden with water vapor that it has absorbed. When it is cooled overhead its moisture-carrying power is reduced and some of this vapor condenses into tiny drops of water or ice, which we see as clouds. If the cooling is continued, more water droplets freeze around the ice crystals until they grow heavy enough to fall to earth as snow, or to melt and become rain. So, with moving air there is also a constant circulation of water, up through the high atmosphere and back again.

But this air, with its burden of moisture, moves over definite pathways before delivering the water back to earth because, as the earth revolves around its axis, the air moves with it, just as the oceans move in unison with the land.

At the equator the earth's circumference is about 25,000 miles. A person living on the equator makes the entire journey around the 25,000-mile circumference every 24 hours. To him the air may seem as quiet as the air inside a speeding train. Actually, he and the air around him are moving from west to east with the earth at a speed of more than 1,000 miles an hour. But if one were to journey from the equator toward the pole, the speed of one's movement from west to east would grow steadily less, as the "circumference" of the earth grew smaller. And the speed of the surrounding air would be reduced in the same way, if there were nothing else to disturb it.

But, fortunately for us, there is. The warmest part of the earth is in the tropics, where the warm earth can give most heat to the surrounding air. So, around the earth at the equator the warmed air expands and rises to float over the cold air above, and then flows out over it north and south toward the poles; while down below, on the surface, the cooler air from north and south flows in toward the equator to fill its place. Thus there is established a vast system of air movement, first toward the equator, then upward and overhead back toward the poles.

As the returning air becomes chilled overhead, most of it drops to earth again before completing the poleward journey. On reaching the earth it will have lost some of its more than 1,000-mile-an-hour eastward motion; but it is still moving a good deal faster than the earth in its new location with the smaller "circumference." Thus, over a wide area of the earth's surface there is created a band of westerly winds, where the air is moving eastward faster than the earth. As this air moves back toward the equator it again reaches the faster moving part of the world, until at last it lags behind the earth's motion and so becomes a northeasterly wind.

The world system of air transportation actually has two forces driving it, therefore. First, the motion from west to east given it by the spin of the earth, and second, the movement between the equator and the poles, given it by the heat of the sun.

This great pattern is complicated by many lesser things, by ocean currents like the Gulf Stream, which carries heat from the tropics and warms the air on the borders of northern Europe; or like the Arctic Current, which absorbs and carries off heat from the coast of New England. The changing seasons have a powerful influence, too. For the earth will change its own temperature very much more quickly than the oceans. Hence, in winter the northern continents become colder than the surrounding oceans and cause the air above them to contract. In summer the land becomes very much warmer than the water, causing the air to expand. These changes have many complicated effects on the air currents and on the weather.

We have seen that where the air comes back to earth before making its return journey to the equator, it is moving faster than the earth, thus making a broad band of west winds. This accounts for the fact that the weather in the United States comes from the west. Here, air moves steadily inland from the Pacific Ocean, laden with moisture that it has absorbed from the water, and then moves eastward across the continent. If there were nothing to chill the air on this journey, there would be no rain. The moisture would be carried out again over the Atlantic and our land would be a desert. This is actually what happens in some places. Studies by the U. S. Weather Bureau have shown that on the average there is actually more moisture in the air over the deserts of southern Arizona than there is over Seattle, Washington. And yet, along the North Pacific coasts the rainfall reaches ten feet a year in some spots, while over the southern deserts it may be one twentieth that amount. The reason for this difference lies chiefly in the contours of the earth.

When the winds blow in from the Pacific, they meet a series of three mountain ranges running north and south to block their path. To continue their journey they must either blow over these mountains or around them. They do both.

First comes the coastal range of California. This is not very high, but to pass it the air is forced upward enough to be chilled and produce some rain. Next comes the valley of central California which, as we would expect, receives very little rainfall. Then the Sierra Nevada range lifts the air higher and receives a good rainfall. After that the air passes over the flat lands of Utah, Nevada, and Eastern Oregon, leaving them nearly as dry as it found them. Finally the

Rocky Mountains give it still a third boost, and draw down another good supply of rain, and again, to the east of the mountains the country goes dry.

Consequently, in the western United States, the land is divided into areas that are well supplied with rainwater, and others that receive very little rain. In these dry areas the land must depend chiefly on the water that comes down from the mountains by way of the streams, and on underground seepage.

And here an almost universal truth becomes especially apparent. *The environment that supports life extends far beyond the vision or experience of the things that live there*. Its most important feature may lie in the distant mountain ranges, perhaps a thousand miles away. And events like forest fires, which affect the ability of the mountains to store and control their supply of water, may decide the issue of life or death for the creatures in the lowlands.

The influences of the mountain watersheds will extend to the farmlands far down the valley of the Mississippi. But here new forces begin to affect the water supply. The streams of air that flowed around the north and south ends of the mountains near the coast have by now begun to draw together again. In the winter the northern air stream sweeps as an icy blast through Canada and then down through the Dakotas. Part of the southern stream flows over the warm Gulf of Mexico, picking up a heavy supply of moisture. When the two streams come together over the central United States, the cold, heavy northern current acts like a barrier mountain range. The warm southern current rises over it, and like the air over the mountains it is chilled to produce rain or snow. Storms of this kind are familiar on every newspaper weather map that shows the meeting of warm and cold fronts. They are an important source of moisture in the central United States in winter.

In warm weather the local thunderstorm takes its place as an important water producer. It comes chiefly as a result of temperature differences on the earth's surface. There may be many causes for these differences. For example, the dark earth of a plowed field will absorb more heat than the surrounding forest, and over this warm field the air will rise. As it goes higher the moisture in the air begins to condense into water droplets, producing the towering cumulus clouds whose contours outline the movements of the rising air. Given the proper combination of heat, moisture, and subsequent chilling, the cloud will at last build up to produce a thunderstorm.

4. *life prepares its own environment*

The wide variations in rainfall over different parts of the country produce important effects on the quality of the soil. We might expect soil fertility to increase with abundant rainfall. But actually it often deteriorates. We have seen how rain combines with carbon dioxide from the air to become a mild acid which dissolves minerals from the rocks. If the rain is not too heavy these dissolved minerals stay near the surface, where the plants can use them for food, and under such conditions the soil may become very fertile. Some of the most fertile soil in the United States, for example, is in the Arizona desert, where these mineral solutions have accumulated for ages. With irrigation, a desert may become fabulously productive. But with too much irrigation, some have been ruined; for when the soil is given more water than it can hold, its dissolved minerals are washed away in a process known as leaching. They are carried out of reach of the plant roots. The water that brought out the productive power of the soil can thus ruin it and turn it back again to desert, minus its former potentialities.

When the land receives too much rain it tends to rob the soil in the same way, leaching away the dissolved minerals. The dividing line between an annual rainfall that builds and one that leaches comes somewhere between twenty and thirty inches, depending partly on the temperature and the amount of moisture evaporation, and partly, too, on the quality of the soil and its ability to absorb water and to hold it. This ability in its turn depends very largely on the help of plants and animals which bring to the soil new qualities not possessed by the original rock particles.

A good soil is really a combination of three basic ingredients. First, the rock particles that are its foundation; second, the organic matter given it by dead plants and animals, and third, a community of living plant and animal organisms.

Air, rock, water, and sunlight—these are the four sources from which come all living things and their environment. On the bare sands of the desert the sun's rays strike in tiny units of energy moving with atomic speed. Some of them we can feel as heat or see as light. These speeding units impart some of their energy to the dead sands, which temporarily store it in the form of heat, but when the sun sinks, this newly acquired energy is radiated back into space and lost. The sand becomes as cold and dead as ever. But chlorophyll in the leaves of green plants exists as an agent for garnering these units of solar energy. It makes of the green leaf a laboratory in which nature creates food for living creatures and carries on unceasingly the magic of building life.

Like the sand, a field of grass absorbs the sun's rays; but when night comes the grass does not give back this newly gained energy. In its green laboratory the chlorophyll blends the sun's captured radiance together with elements taken from the air, the water, and the soil, and builds these dead materials into organized living form to make new blades of grass.

This grass is cool and quiet, giving no hint of the sunlight stored within its framework. But dry it out and touch a match to it. The blades of grass—these tiny bits of organized gas and sunlight—blaze up with a flame hot enough to kill a man. All of that fierce heat is merely a release of the same energy that the cool, moist plants have been quietly gathering from the sunlight and storing for later use.

If the grass is not burned, the energy will remain stored within its substance. If it is eaten by an animal, its life force is transferred with it into the body of the animal to sustain the spark that we call life.

But, for all its abilities as a life builder, the plant is not entirely self-sufficient. Its principal building material is carbon, which it takes from the air. This carbon, once it is locked up in the plant, becomes useless for other plants unless it is later released into the air again. Through the ages that they have existed on this earth, plants have locked up enormous quantities of carbon in this manner and stored it in the ground often in the form of coal. If we could picture the thousands of tons of coal that are used every day throughout the world and could realize that, with all that has been burned, there is enough left to last us for another thousand years or more, we could appreciate something of the amount of carbon that has been withdrawn from the air. But even this is only a drop compared with the amount that is used by plants and then returned to the air again by the animals and bacteria that eat the plants, decompose them, and release their elements. Without the help of this cycle of use and return, the plants might before now have eliminated all life from the earth, by using up all of their basic food.

Some scientific evidence seems to indicate that in the early days of the world there was not enough free oxygen available in the air to support animal life. But there is oxygen locked up in the plant's two chief foods, carbon dioxide and water. From these the plants draw, respectively, carbon and hydrogen, and in the process they release oxygen in a form that animals can use. Through this action of vegetation the world's supply of free oxygen has been built and maintained.

So, even before our environment could start to support us, it had to be built up by a multitude of interacting forces, including living things. And the hierarchy of life had to start from the beginning with the simplest microscopic single cells which, step by step, prepared the way for higher forms.

There exist dramatic examples of land where man has destroyed all life. In them we can see some of the futile attempts of our common plants to make a fresh start, and can visualize something of the slow process by which an environment must build up from its simplest beginnings before it can support life as we know it.

5. *life comes to the soil*

There is a spot in the woodlands of southeastern Tennessee that can never be forgotten by one who has seen it. To reach it, one may travel for a hundred miles through forest-covered hills, rich with laurel, azalea, and rhododendron, and along springs and brooks and ravines which sometimes open up into green meadows where cattle graze.

Suddenly this green world disappears. The forest gives way to a hundred square miles of desert as dead as the Sahara. The rolling hills are cut into rows of low, steep-sided ridges, sterile and bare of any life. The soil is dry, the springs and brooks are gone. In this area the annual rainfall is less than in the surrounding country. The winds are stronger. It is hotter in summer and colder in winter. Here and there on this desert there stand in rows the dead skeletons of small trees, planted by people who hoped to start a new forest.

The soil in the nearby woodland is dark, rich, and spongelike. That on the desert is coarse, hard, and yellow. This desert was once covered by a forest and by rich forest soil. But today that soil lies five miles down the valley at the bottom of a reservoir and the shoals of coarse desert soil grow deeper, year by year, as every rain washes its fresh quota down to the reservoir.

This all happened because, many years ago, a copper smelter was built here, and the fumes from the smelter killed the surrounding trees, thereby setting in motion a train of events that finally produced the desert. The owners of the smelter have long since learned to control these fumes, which no longer poison the air so seriously; but the harm has been done.

After the fumes were controlled, many attempts were made to restore the forest. Desert grasses were planted, in the hope of furnishing some green cover to hold the soil in place; for rich soil cannot exist without the help of plants to build and protect it. But with the killing of the forest, the living soil that gave it life had also been killed. There were no roots to hold the soil in place, no litter to absorb the rain, and the grass seeds washed away, the seedling trees withered, and the dead soil continues to this day to wash from the desert and drift down to fill the reservoir.

At prohibitive cost, this desert land could be restored to life. Modern machinery might partly substitute for some of nature's processes. It could fill the eroding gullies and build level ridges along the hillsides to stabilize the soil, hold the rain, and so give moisture to the earth. Then seeds of grass and trees could take root and find a chance for life. Given such a start, nature could take over and slowly rebuild the ruined soil, organizing once again the community of living things that makes life possible for the forest. This might even be accomplished in a human lifetime.

Without man's help, nature might take ten times as long, for her processes are slower. Life cannot make a new start here until the earth offers it a firm foothold, and this must wait until nature has washed away the dead soil down to a solid foundation. This is the process that we call erosion. In this particular area it has been going on for many years and the land still looks as dead as the excavations for any city building. Eventually nature will find a solid footing, whether it be a ledge or a rock heavy enough to withstand the movement of the soil. Here seeds may find shelter for a foothold.

Or life may even start on the face of the rock itself. The rock offers little in the way of food and moisture, but on its secure base the pioneer plants may gain a foothold. Such plants must have the ability to dry up and lie dormant through times of dry weather, then waken to absorb the moisture of every rain or heavy dew. Many species of lichen have this power. Their rootlike fibres secrete an acid which dissolves minerals from the rock. Eating their way into it, they prepare an entrance for moisture which may later freeze and crack off rock particles— the beginnings of soil. The lichen thus offers to other more delicate plants a seedbed with moisture, a foothold on the rock, and mineral solutions for food.

In this seedbed, mosses, annual "weeds," and hardy ferns may grow and die, adding their substance to it, catching wind-blown dust, and building a deeper bed where seedling trees and other plants may find food and moisture. The growing community of plants will slowly spread over the rock till at last the roots reach into the soil around it. Here the tiny hairs will fill the earth so compactly that they may touch every soil particle, tying it firmly into place, making a secure foundation for a further spread of new plants around the parent rock.

A root system is a really incredible thing. Many studies have been made of its extent. In one study, a plant of winter rye grass was grown for four months in a box with less than two cubic feet of earth. In that time the plant grew twenty inches high, with about 51 square feet of surface above the ground. But underground the root system had developed 378 miles of roots and an additional 6,000 miles of root hairs! This meant an average growth of three miles of roots and 50 miles of root hairs for each day of the four-month growing season. The growth rate varies with different plants, of course, but this gives us some idea of the activity that goes on under the surface of a quiet-looking meadow, while the grass prepares food that will later become milk and meat and butter for us.

But these growing roots are doing far more than just binding together the rock particles that form the soil. They are taking the first step toward creating an entirely new kind of soil.

A Flemish physician who lived in the 17th century gave an interesting picture of this when he tried growing a willow sprout in a tub of earth. For five years nothing was added except rainwater, and the willow grew into a small tree. At the end of the five years the tree was weighed. It had gained more than 164 pounds in weight, while the soil in the tub had lost only two ounces. Actually the soil weight must by now have included millions of microscopic root hairs from the tree, but the figures are accurate enough to show that those 164 pounds of tree must have come from somewhere outside the soil.

If we divide up the plant into the various elements that form its substance, we find that only five percent of its weight comes originally from the soil. The elements in a mature corn plant include carbon, 44.58 percent; oxygen, 43.79 percent; hydrogen, 6.26 percent; nitrogen, 1.43 percent. These all come originally from the air and water, and together they form more than 95 percent of the plant. A good proportion of them comes to the plant through the roots, by way of the soil, after earlier plants have fitted the soil to receive them. The rest of the plant, that which comes from the soil itself, includes potassium, 1.62 percent;

calcium, .59 percent; silicon, .54 percent; magnesium, .44 percent; phosphorus, .25 percent; chlorine, .20 percent; sodium, .15 percent; iron, .10 percent, and sulphur, .05 percent. Some plants contain very small amounts of other elements, such as copper, boron, and cobalt. These we call trace elements.

All these elements are built together into a living plant through the agency of the chlorophyll, the green coloring matter that is carried in the leaves. This building process is the essential first step that prepares the way for all the life that exists on earth. Chlorophyll has not yet yielded to man all the secrets, either of its composition or of the magic by which it transforms inert building blocks into living material, but we do know certain basic facts. The essential first step consists of building sugar out of sunlight, carbon dioxide and water. This is called photosynthesis (putting together). To make one molecule of sugar the chlorophyll produces the union of six molecules of water and six of carbon dioxide. With them it binds the energy from the sunlight, and in the process six molecules of free oxygen taken from the water and the carbon dioxide are released into the air.

As the roots spread through the soil, they fill it with this new living substance built from air, sunlight, and water. But this has not yet become a part of the soil. It is not until the plant itself dies that the dramatic change takes place in the soil. For now the dead plant's roots and leaves offer food to the small organisms that are among the most important factors in the whole cycle of life, the bacteria, the molds, and the rest, most of them too small for the eye to see. Their most important function lies in decomposing the remains of the higher plants and animals, changing them into new chemical combinations that can be used again by succeeding plant generations for food.

The decomposing plant attracts a host of small creatures that help to break it up. Earthworms eat it, mixing it with the soil particles that pass through their bodies, digesting the whole, and casting it up on the surface, a revitalized and richer soil. The number of earthworms in the soil depends largely on its chemistry and on the amount of plant material they find in the earth.

Myriads of small creatures spend parts of their lives in the soil: ants, beetles, wasps, spiders, and many others. Some of these come to eat the plants, and many meat eaters come to eat the plant eaters. Among these the shrews and moles play a very important part. In favorable locations there may be as many as 100 shrews to the acre, and each shrew may eat the equivalent of its own weight in other living things each day. All this activity combines to carry on the work of plowing, mixing, and fertilizing as the creatures add their remains to the land.

This hive of living things in the soil, the eaters and the eaten, adds up to incredible numbers. The bacteria alone may range from comparatively few up to three or four billion in a single gram of dry soil. At the Rothamsted Experiment Station in England it has been estimated that in good soil the bacterial matter, living and dead, may weigh as much as 5,600 pounds per acre. Does this seem like a small amount in a whole acre? At the rate of even one billion to a gram of soil the total body surface of the bacteria in an acre, if spread out flat, would equal 460 acres.

The fungi may add up to a million in a gram of dry soil, weighing over 1,000 pounds to the acre.

Each of these small living things adds its tiny bit to the building of the living earth until, in the average

acre of good topsoil, with four-percent organic matter, there are stored about 80,000 pounds of such
organic matter from plants and animals, containing
energy from the sunlight equal to that in 20 to 25
tons of anthracite coal.

But while the soil lives, this stored-up energy is
constantly being used for food by the teeming life
it supports and, as we have seen, it must be constantly
renewed by the plants in order to maintain this life.
For good soil is actually a living thing, and its health
is a matter of life and death to the plants and animals
that live on its surface. We ourselves are as dependent on its health as the smallest of its creatures.

6. *the soil responds*

As the plant roots and fungi grow into the soil, tying its rock particles together into a firm mass, opening the way for other living things, and filling the earth with organic matter containing packaged energy from the sunlight, a subtle change occurs.

The root tips release carbon dioxide, the source of carbonic acid. This reinforces the action of the rain, helping to dissolve minerals from the soil particles, making them available to the plants for food. As the bacteria decompose the dead plant matter, they too release carbon dioxide and contribute their share toward the enrichment of the soil. The decomposition of animal and vegetable residues by microorganisms produces many other acids, besides carbonic acid, including citric, tartaric, oxalic, and malic. Such acids are probably of even greater importance in making minerals available.

In raw, unprepared soil these dissolved minerals might be carried away by every heavy rain, and the supply left for the plants would be rather precarious. But here the bacteria provide another service, for in the decomposition process there are some parts of the dead plants and animals that are more resistant than others. These stay in the soil for a long time, forming a dark, spongy, very absorbent material called humus. Humus stores rainwater, with its dissolved minerals, holding both as a reservoir for plants to draw on. It also stores minerals drawn up from below by the deeper roots.

As the bacteria use up their food supply, billions of them die of hunger or become inactive, and the life process in the soil slows down until further stores of food are added by plants or animals. As in most of nature's activities, this whole life cycle in the soil becomes a self-regulating system—an organized community, adjusting its numbers to the food supply so long as it is undisturbed by outside forces.

On the surface this community may appear to be merely a blanket of dead leaves and litter from last season, but under these lie the decaying remains of their predecessors from earlier seasons. As we go deeper these become mixed with soil particles and with the roots of plants living and dead. Through this material run myriads of passageways left by insects, mammals, and decayed roots. The whole makes a perfect protection for the earth and a sponge to check the runoff of rain, which it absorbs and introduces slowly into the soil reservoir below.

This power of the organized topsoil to store water and minerals is the key to the next step for the developing plant community.

For, while minerals form a very small part of the whole plant—only one twentieth—that small fraction, together with sunlight, is the key that makes the whole function.

This has been proved in some very interesting tests in Kansas. In the western part of that state, where there is comparatively low rainfall, the good earth is able to store most of the rain and the dissolved minerals near the surface. But in the eastern part of the state the heavier rainfall supplies more water than the soil can hold. Here a good share of the mineral solutions is leached away. In a comparative analysis of wheat raised in the eastern and western parts of the state, it was found that grain raised in the dry western part contained nearly 50 percent more protein than the same kind of wheat raised in the eastern part. An important cause of this difference appears to lie in the leaching of the eastern soil which, having lost a share of its dissolved minerals, was unable to give the wheat a proper supply of them.

Obviously the quality of the soil has a great effect on the quality of the plant. And since animals, including man, require roughly the same elements as plants, the health of man and the lower animals must depend on the ability of the soil to supply those needed elements in the right proportions.

Some interesting experiments have been performed to show this interrelationship. At the University of Missouri, Dr. William A. Albrecht grew hay on three plots of ground which were somewhat deficient in minerals. One of these plots was given no fertilizer. It produced about 1,700 pounds of hay per acre. The second plot was fertilized with nitrogen. It produced nearly double the crop of hay—more than 3,200 pounds. But when fed to rabbits, a pound of this nitrogen-fed hay produced less rabbit meat per pound than the much less luxuriant unfertilized hay. The nitrogen had forced its growth, but the soil had been unable to supply enough of the other minerals to keep pace with it.

The third plot was given a balanced fertilizer with the essential minerals that the grass needed. This produced less hay than the nitrogen-fed plot—about 2,400 pounds—but when fed to rabbits it produced nearly twice as much meat per pound of hay. So it appears that the minerals can be of greatest value only when the plant can use them and when they are in the right proportions. The actual figures shown in these tests were:

Unfertilized hay	23 lbs. of hay produced 32 oz. of rabbit growth.
Nitrogen-fertilized hay	23 lbs. of hay produced 26 oz. of rabbit growth.
With balanced fertilizer	23 lbs. of hay produced 49 oz. of rabbit growth.

From all this it might seem that plants grow best in a land where there is little rain. Actually, most food plants need huge amounts of water. For example, a single corn plant uses about 50 gallons in 100 days of growth. In one test made in Kansas, an acre of corn containing 6,000 plants used 325,000 gallons of water in 100 growing days. This equaled the amount of water it would have taken to cover the acre 11 inches deep—or 11 acre inches.

To demonstrate the value of water, in an experiment in Utah some corn was raised on dry land without irrigation. It produced 26 bushels per acre. On another plot the corn was irrigated with 15 acre inches of water. It produced 53 bushels to the acre, more than double the other crop.

Hence, to raise a good crop of plants, the land must carry an adequate supply of moisture. It must have the ability to store it without leaching. To do so it must be well supplied with humus built into it by the plants and small living creatures of the soil. It is commonly estimated that nature may normally take as much as 500 years to build an inch of topsoil. This topsoil is one of the keys to man's existence on earth.

7. *the forest grows*

As the tree draws in its raw materials from the air, water, soil, and sunlight, these are carried to the leaves. Here, through the miracle of chlorophyll, they are woven together and transformed into sugar. And, from this sugar, into the vast number of chemical combinations that form the living substance of the tree—into roots and leaves and branches, into flowers with their male and female parts that together produce seeds. In the seeds are stored food and microscopic cells for the production of new life. The cells will later develop, each one producing its own assigned part of a new plant, with vitamins, enzymes, and other chemical combinations that interact to make it function as a living thing.

Some soil fungi form a direct partnership with trees and other plants. Beeches and pines apparently cannot make healthy growth unless there exists an active association between their roots and certain kinds of fungi. This partnership—or symbiotic association—is known as mycorrhiza. Its exact function is something of a mystery, but it evidently plays an important part in the transfer of food from the soil to the root system.

One group of plants, the legumes—clovers, beans, locust trees, and other pod bearers—joins forces with bacteria to form a sort of chemical laboratory in the earth. When nature has built up the soil's chemistry to a condition that will support them, these legumes take their place in the plant community. They offer a home in the soil to nitrogen-fixing bacteria which enter their roots and cause them to swell into lumps called nodules, where the bacteria live in colonies of many millions. Taking their energy from the sugar in the plant roots, the bacteria gather nitrogen from the air to form nitrogen compounds, which they store in the nodules.

When the roots die, the nitrogen is left in the soil, and with this enrichment the plant community bursts into full life.

Near Athens, Ohio, a plantation of cedar trees was set out on an area of very poor soil by the U. S. Forest Service, who wanted to test different trees in an effort to start a successful forest. In one part of the cedar plantation they set out among the cedars a number of locust trees (legumes), which carried the nitrogen-fixing bacteria on their roots. Eleven years later the cedar trees, planted alone, stood on the average about thirty inches high, while those among the locusts averaged perhaps seven feet. Between the small cedars the ground carried a thin, sickly cover of low grass, an occasional white poplar, and very few other plants. The soil was dry and offered little food or shelter for living creatures. Here nature might need a century or more to establish a healthy forest.

Under the locust-and-cedar combination, however, one can step across from a semi-desert into a rich young embryo natural forest. The ground is covered with a lush growth of grasses, weeds, and vines. On its surface a litter of dead leaves and stems has begun to collect from plants of past seasons. Beneath the litter the soil is cool and damp with moisture stored from the rains. Shaded from the strong sunlight, it is a good natural bed for the seeds of forest trees to make their start in. And here are growing seedlings of tulip trees, red oak, red maple, white ash, and others.

Both groups of trees were given an equal start on the same kind of soil, and the entire area has been bombarded every year by millions of seeds of many kinds, brought in by wind and bird and mammal.

Why are the white poplars the only new trees to start among the cedars? Why are there so few poplars under the taller cedar-and-locust combination, while many other trees have started here whose seeds failed to survive out in the open? The answer is that each plant is a specialist, adapted by its own habit of growth and its own special requirement for light and moisture to grow best in its own preferred environment.

Among the cedars only the seedlings of the grasses and the poplars have been able to withstand the drying heat of the strong sunlight. Under the taller, richer, cedar-and-locust growth the white poplars have had to meet the competition of red oak, red maple, white ash, and tulip, whose seedlings thrive in partial shade.

These trees are growing here, not because the ground received any more of their seed, but because the seeds that fell found the added moisture, soil quality, and just the amount of protection from sunlight that they must have to get a secure start in life.

This young forest will now follow the normal steps of forest development; for the seedlings will eventually grow up to overtop the earlier trees that nursed them and finally crowd them out by robbing them of sunlight and the free space they need for growth. And with the early trees will go many of the small plants of the ground surface which also need sunlight.

For many years this slow development will continue, until the crowding newcomers start to battle among themselves. At length their towering crowns will cut off the sunlight that their own seedlings must have in order to live. Then they will have destroyed their own power to reproduce. But among the many kinds of seedlings now struggling in the forest shadows, there are some that do thrive under these new conditions. Chief among them are the hemlock, beech, and sugar maple. These will now outgrow the others in their race to reach upward for the sunlight, finally touching their tops against the forest canopy. Here they will await their turn until a windstorm strikes some giant oak or tulip to the ground, making room for one of the newcomers. And at last these will take over to form the enduring climax forest. They are now the dominant trees. Nothing else can compete with them under the conditions they have established.

In some forests, where the soil is shallow, as the great trees fall, one by one, tearing their spreading root systems out of the shallow soil, the roots lift huge balls of earth with them. In this process the ground is plowed up into series of small ridges and depressions that fill with water after rains or melting snow. The water is held until it sinks slowly into the ground, and thus is reinforced the absorption system already established by humus and insects.

The trees have even made a new climate for all the lives that exist under them. Near Cleveland, Ohio, measurements were taken to compare the climate in a climax forest of this type with that in an open field nearby. It was found that on a bright day there was 750 times more light in the open field than in the forest. The trees slowed down the speed of the wind until, 1,000 feet inside the forest border, its velocity in summer was only one tenth as much as in the open field. In winter, when the leaves that slowed it had fallen, wind speed increased to one quarter that in the field.

These differences influence the moisture in the forest; for the wind and the sun's heat have a great effect on evaporation. It was found in the Cleveland tests that average moisture evaporation was 55 percent less in the forest than in the field during the summer, and 38 percent less in the winter.

With its blanket of humus and its protection from chilling winds, the forest soil is guarded against freezing in the winter, while the branches shade the snow against rapid melting. The soil is always ready to receive the water from the slowly melting snow, while the surface of the hard-frozen field nearby sheds the snow water like a roof, as it floods off under the warmth of the spring sun.

So, in addition to building timber and providing environment for many kinds of plants and animals, the forest also builds a reservoir to catch and store the huge amounts of water it must have for growth. From this reservoir, springs break out and brooks slowly cut their channels and bring life-giving water to land outside the forest.

8. *the forest community*

Our growing community of plants now provides three essentials needed by other forms of life—shelter, water, and a dependable supply of food. Many forms of plant eaters now come to use it—insects, mammals, birds, and other creatures.

Here, for example, is a caterpillar busily transforming the organic substance of a leaf into the juices and organs that make up the parts of a caterpillar. Eventually this substance will go through a series of further changes until at last it turns into a flying insect. Then it will repay with interest the damage it has done to the plant, for it becomes a partner in the plant's life process, carrying pollen to fertilize the blossoms. Insects make possible the continued existence of many plants.

Nearly all the fruits and vegetables used by man are directly dependent on this partnership with insects. But this partnership requires the most exacting regulation to meet two fundamental laws of nature on which all life is based.

First, insects, like all living things, must have the power to multiply faster than their normal death rate to insure against the catastrophes of disease and weather. Without this insurance no species could survive.

But this power carries with it tremendous danger; for insects, if allowed to multiply unchecked, would soon destroy all the leaves and kill all the plants that support them. We can see that danger most clearly in the life of soil bacteria, the smallest and simplest of all living things. One of these invisible cells may seem like a very insignificant part of the living community. But watch its numbers grow.

Each individual multiplies by dividing into two complete new ones. Under favorable conditions this may happen about twice each hour. Even if it happened but once each hour, and if each one lived, the offspring from a single individual would number 17,-000,000 in 24 hours. By the end of six days the 17,-000,000 would have increased to a bulk larger than the earth, and every living thing on earth would have become engaged in a suffocating struggle for food and air and life.

The useful bacteria that, kept in their place, make possible the higher forms of life, would have turned into irresistible destroyers.

The same principle, of course, applies to all forms of life. The insects that pollinate the blossoms are no exception.

To control them, nature uses a highly organized police force of flesh eaters—bacteria, insects, mammals, and birds, each one a specialist designed for a particular role to which it is best adapted.

For example, in this small community we are watching, the larvae of many insects spend parts of their lives in the upper soil. Shrews hunt them under the leaf mold. Other insects and some of the molds and bacteria also feed on them. On the surface many ground-nesting birds, such as the towhee, turn up the leaves to find them. The brown thrasher hunts here too, and continues the search among the bushes where it makes its nest. It is joined by the warblers and vireos, which extend their search up to the tree tops.

As the trees grow larger and their lower branches die, the fungi may decompose and soften the wood in the knotholes, offering favorable nesting sites for woodpeckers and nuthatches, whose feet are adapted for hunting on the bark of tree trunks. The woodpeckers go one step further, drilling holes through the bark to catch the insects hidden within.

Some of the insect pupae that survive this search emerge to fly over the tree tops, but here they are met by the swifts and swallows by day, by the nighthawks at dusk, and later by the bats that are equipped with nature's radar systems to hunt in the darkness. Each one of these controlling predators must in turn hold its own place against others that are larger or stronger or more active—hawks, owls, foxes, weasels.

It may seem a haphazard collection of predators and prey. Actually it has evolved into a very highly organized, beautifully controlled form of community government, with its own automatic system of farming, pedigree breeding, sanitation, policing, and insurance.

There may be many niches in the forest that give shelter for a nest or a den, but their inhabitants can survive only if the surrounding territory can offer enough food to support life. They can rear their families and multiply only if they can find the added food to feed their young, and many young birds and animals require the equivalent of their own weight in meat or insects each day.

As the growing families move out to forage for themselves, they find the best hiding places occupied, the best feeding areas already in use. They are then more exposed to their enemies. The first to tire or weaken in the search for food become the easiest prey. In this way nature removes the unlucky and the weaklings, saving in the long average the best, hardiest, or quickest to learn, to live and to carry on their kind. Thus their population is adjusted to the number that the land can safely support.

Sometimes a predator species may be wiped out

for a time by weather or disease. But nature carries insurance against such catastrophes, for the area patrolled by each species will slightly overlap that of its neighbors. For example, if the ground-dwelling towhee disappears, the thrasher, the shrew, and the white-footed mouse all are ready to step in to reap the more abundant food supply and, with more plentiful food, to rear larger families.

This local police organization is backed up by a more mobile patrol force; for the larger predators, requiring more food, must manage to cover more ground than their smaller prey, taking off the season's increase wherever it is easiest to catch. They concentrate on areas of heaviest production.

In this hierarchy of life some creatures may fall prey to smaller predators, the disease germs that attack them from within, or the parasites from without: the worms that live in their bodies, for example, or the fleas and lice living on their skins. As a rule, these take only a small share of the daily product of energy from their host, for the body of a healthy animal can usually control such parasites. But when the host is weakened by hunger or age, the parasites may prove fatal. Then the sanitary agents of the community step in to do their share. Vultures eat the dead flesh, bacteria and the larvae of flies feed in it and decompose it. Burying beetles carry its remains underground to enrich the soil, laying the foundation for a new cycle of life on a cleansed surface.

So the living community in the forest grows, the plants drawing in elements from air and water, and energy from the sun, building them into life—the plant eaters staying near their stationary food supply, turning its energy into meat, passing this on to the larger, more mobile flesh eaters. These in turn pass on their energy to the hierarchy of larger forms, each group in its turn becoming fewer in number as the larger creatures require more small bodies to feed them, with each successive individual forced to cover the larger area of ground needed to raise these many smaller lives. But in the end even the largest succumb to the bacteria and beetles that complete the circle in the earth.

All these creatures live and flourish completely unaware that they owe their existence to a few hundred million invisible bacteria living in the roots of some locust trees which pioneered the way for them many years ago.

9. *the community through the seasons*

As we look at the quiet stillness of a forest we may imagine something of the feverish activity going on within it, and some of this we may see if we have the patience to watch for it.

Throughout the ages of its existence the fortunes of the great forest community are constantly fluctuating under the influence of four fundamental forces and a host of lesser ones. The long-range cycles of weather, the cycles of disease and prosperity among member groups, and the resulting changes in the influence of these groups on each other, all play a major part in the existence of a whole community. But the changing seasons of the year have the most obvious effect.

In winter the plants of the northern climates rest from their function of producing food, and all the creatures of the forest must adjust their lives to a reduced food supply. Most of the birds go south. Many animals, such as the chipmunks, woodchucks, skunks, snakes, and frogs, and many insects, retire for a long sleep during the winter. Those that remain active must depend on food stored up from the summer's harvest.

The ruffed grouse and cardinal will find it stored in the seeds, fruits, and buds of trees and bushes. The nuthatch finds the eggs of insects hidden in the bark of trees. Woodpeckers drill through the bark to find the insect larvae that winter beneath it. The red squirrel, flying squirrel, and gray squirrel live on stored seeds and nuts. Perhaps the most active and numerous animals of the forest are the white-footed mice and the shrews, and in them we see a good example of interdependence. The mice live on their stores of seeds and on what insects they can find hidden in winter retreats, while the shrews hunt day and night, digging tunnels through soil and humus and through rotting logs to capture the mice, as well as great numbers of insects.

These mammals and birds in their turn offer food to the fox, the weasel, and the barred owl, which of course can exist only in smaller numbers, since for their support they require so many of the lesser creatures.

As the snow melts and the suns of March and April warm the ground, a change comes over the forest. The low plants on the earth's surface come to life. The early insects come out of hiding and many more hatch to take advantage of the new supply of food. The skunk, the chipmunk, and several kinds of snakes, toads, and frogs wake from their winter sleep, and all include in their varied diets great numbers of insects. Now the ground grows bright with spring flowers—spring beauties, yellow adder's-tongue, hepatica, trillium, and many others. The flower buds of the elms and red maples offer food for gray squirrels. Frogs add their music to the notes of birds returning from their winter in the south. For now nature's food factory has begun again to build new life from sunlight, air, and water. The green algae in the pools, the forest plants, shrubs, and trees are all doing their share, offering their stores of energy to the creatures that come to feed on them. And the creatures respond. Insects of different kinds attack every part of every tree and plant—buds and blossoms, leaves and bark and wood. Spiders and predatory insects feed on these plant eaters, and all are in turn preyed on by the larger animals, including snakes and frogs, and by waves of migrating birds that spread up from the South to find nesting sites and hunting grounds to support their hungry young.

Many of these birds stop merely for a rest and a meal, and hurry on to find summer homes farther north. Others stay to fill every available niche in the forest and, as the leaves unfold, the branches that yesterday were bare and inhospitable, now offer shelter from weather and predators.

Dr. Arthur B. Williams of Cleveland made an interesting four-year study of the relationships of the nesting birds in a 65-acre tract of climax beech maple forest near his city. During these four years the number of pairs of nesting birds present on the tract were 136, 174, 176, and 134. This gave a yearly average of 2.3 pairs of nesting birds per acre. When one thinks of the enormous number of insects needed by a growing family of young birds, each one requiring nearly its own weight in them every day, this seems like a very small territory to supply the needs of each family. But actually the territory includes more living space than might appear, for it reaches upward to the tree tops and above, as well as along the ground. Within this territory, as we have already seen, there are many different kinds of hunting grounds. Each different kind of bird is specially adapted to hunt in its own niche, and each, having selected its own hunting territory, will defend that territory against all competitors of its own species. But many different kinds of birds may nest close together without competing, because each species occupies a different niche, while others that might compete are so scattered as to cause little trouble.

For example, Dr. Williams found four different kinds of woodpeckers nesting in the area, the pileated, hairy, downy, and red-bellied. But since each species is so specialized as to confine its hunting to a different part of the larger trees, there was apparently little competition among them. Three pairs of white-breasted nuthatches in the area might have competed with the downy woodpeckers in hunting among the bark crevices, but as their nests were widely separated, there was probably no serious competition. The chickadees and tufted titmice perhaps competed with the nuthatches, but their nests also were well scattered, so that the area allowed about 3.6 acres for each pair.

The flycatchers, living on flying insects, divided the territory, each pair of Acadian flycatchers having staked out its own separate nesting and hunting ground in a little glen by a stream. The crested flycatcher used a different part of the woods, and the wood pewees used widely separated territories in the higher areas of the forest.

On the ground and among the lower bushes, the Louisiana water thrush, wood thrush, cardinal, ovenbird, and towhee shared the territory, which totaled about two acres of hunting ground for each nesting pair. The hooded warblers and redstarts hunted chiefly at a higher level. In the leafy foliage of the taller trees the tanagers, vireos, and cerulean warblers were the chief hunters, while the blackthroated green warblers confined themselves chiefly to the upper branches of hemlocks.

Thus, every part of every tree, bush, and plant in the forest received its regular protection from its own special guardian.

As the young birds grew and left their nests, the

hunting territories were less strictly guarded by their parents and finally were abandoned altogether as great numbers of young searched for their own food. In the process of growing up, many young died of accident and exposure to weather and by predator.

By summer's end, conditions began to change again for our forest community. Much of the tree and plant growth had stopped, providing less appetizing food for insects. Their own time of greatest multiplication had passed. The birds begin to move south again. The redstarts are among the first to go, followed closely by the ovenbirds. By the end of August the red-eyed vireos and wood thrushes have gone. Purple martins and chimney swifts busily hunt flying insects that seek sunshine above the dark canopy of the forest.

By the beginning of October the scarlet tanagers and hooded warblers, last of the summer birds, have gone, and now the forest is filled by great waves of southward-bound migrants coming down from the north. Hundreds of robins, thrushes of several kinds —first the hermits, then the olive-backs, veeries, and gray-cheeks stop to feed on wild grapes and on the fruits and berries of many trees and bushes. They vary this diet with beetles, grubs, and other insects, which they dig out from under dead leaves.

By early November most of the leaves have fallen and the bare tree tops again admit light to the forest floor. The robins and thrushes continue southward. Bobwhites may come in from the fields to gather beechnuts in the woods. Tracks in the early snows of late November tell of the search for food by squirrels, cottontail rabbits, red fox, white-footed mouse, and short-tailed shrew.

So the population of the forest fluctuates greatly throughout the seasons. Dr. Williams estimated the

average bird population for the 65 acres that he studied to be as follows:

	March	May	July	Sept.	Oct.	Dec.	Jan.
Permanent residents							
woodpeckers							
barred owls							
titmice, chickadees							
cardinals, nuthatches							
towhees, etc.							
total 11 species	100	69	52	78	86	106	102
Summer residents							
vireos, thrushes							
warblers, robins							
flycatchers, tanagers, etc.							
total 10 species	0	365	184	195	510	2	2
Autumn and winter visitors and transients							
red-breasted nuthatches							
juncoes, thrushes							
warblers, sparrows, etc.							
total 56 species	200	505	3	295	465	218	50
Grand total	300	939	239	568	1061	326	154

All this great company of birds, mammals, and insects in Dr. Williams's study must adapt itself to the environment established and controlled by three kinds of trees—the beech, the sugar maple, and the hemlock. The tulip, red oak, red maple, and white ash play important, but secondary, roles. The chestnut once shared a dominant position until it was wiped out by blight.

With these seven trees there are 20 other lesser kinds that grow in the forest openings or borders or under the shade of the dominant trees. With them are seven species of vines, 45 common herbs and shrubs, together with 24 other rare or uncommon species; 14 ferns, 10 mosses, and six miscellaneous algae, lichen, liverwort, and sedge. Besides these, it is estimated there are between 1,200 and 1,500 different species of large fungi and other plants that live on dead organic matter from trees and plants.

Dr. Williams concluded after his study that the existence of this great forest depends on the birds, mammals, and predatory insects that protect it from its enormous population of plant-eating insects. On the other hand, the insectivorous creatures must have insects to feed them, and many of the insects, as we have seen, play an important role in preparing the forest soil and pollinating plants.

Thus, in every forest the living creatures that make up the community are actually selected by the dominant trees and the lesser plants that determine the environment in which they must live.

From all this we see that the forest is a great organization made up of many separate and indispensable parts. Some of these parts may appear to be harmful to its life. But in most cases the degree of harm or value will depend on the perfection of the control or balance that the different members achieve among themselves.

Owing to the hazards of climate and disease, this balance is never quite achieved, and its fluctuations play an important part in forest life. And on the degree of its attainment will depend the amount of life that the land can support, in other words its carrying capacity.

It is interesting to note how this principle is applied by nesting birds in the forest where, as we have just seen, each pair selects and defends enough territory to support its family. But this defense is exerted only against members of the same species, while nests of other, non-competing species might be tolerated in the same tree.

10. community life cycles

The influence of the seasons on the life of the forest community is fairly regular and predictable.

The longer range cycles in its life are much more complex; they revolve chiefly about the fluctuations in climate and in the relations between the living community and its food supply.

Returning to the study of the forest near Cleveland, Dr. Williams noted that there were heavy crops of beechnuts and sugar-maple seeds in the autumns of 1930 and 1931. These are the basic foods that support mice, chipmunks, and gray squirrels, and so they play an important part in the support of the shrews that feed on the mice. In the fall of 1931 the forest was filled to capacity by shrews, mice, chipmunks, and gray squirrels.

However, many types of trees go through periods of rest, when they produce very few seeds, or none, and these resting periods are often affected by drought or freezes or other changes in weather. Normally, if the crop of one species fails, there will be crops from other trees to carry the animals through the lean year. But in the fall of 1932 the crops of both beech and sugar maple in this forest were almost complete failures. The chipmunks had little food to store for the winter, and most of them retired to their dens a month ahead of their usual time in late November.

The year before, it had been possible to count as many as 50 chipmunks in a single morning, but this year none were seen above ground. By spring there were very few left alive in the woods, most of them apparently having died through the winter. The more adaptable red squirrels turned to the seeds of the tulip, hemlock, and cucumber trees. But the gray squirrels, which had numbered about 50 the year before, left the forest in a body. Only one individual stayed behind to live through the famine, apparently depending on the scarce acorns, hickory nuts, tulip seeds, and mushrooms.

In the fall of 1933, when a census was made of the white-footed mice and shrews by trapping, it was found that the mice had decreased by 80 percent and the shrews by 65 percent from their numbers of the season before.

This count was complicated by the fact that the summer of 1933 was very dry, and many of the insects living in the humus either died or became dormant, producing no crop of young. Ordinarily these insects are an important part of the shrew's food supply and act as a so-called buffer food to protect the mice. But as the insects disappeared the shrews were forced to depend more heavily on the mice. So at first these actually suffered more loss than the shrews, although their own food supply had not been so seriously affected. But the reduction of both insects and mice now began to affect the shrews, and the following year, 1934, the shrews dropped 50 percent, while the mice showed only a 20 per cent reduction from the 1933 count.

With good food crops in 1934 and 1935, the small animals recovered fast. The chipmunks were completely restored, and the mice increased until, for the first time, they outnumbered the shrews; and now, also for the first time, two new species of mice came into the forest. The pine mouse became common, and the meadow mouse, which does not normally live in the forest, also appeared, perhaps driven in by the pressure of increasing numbers in the fields nearby. The shrews made a good recovery, but they were slower, since they undoubtedly had to wait for the return of the mice that supplied their food.

We may wonder how one could count the shrews and mice in a forest. Dr. Williams made his census by trapping them on five plots of representative land, each about ten yards square. While there was, no doubt, considerable travel back and forth across the boundaries of the plots by the animals, and the counts may have included some outsiders, they still give a good comparison of numbers. The total for the five plots showed:

	1932	1933	1934	1935
Short-tailed shrew	37	12	6	23
White-footed mouse	28	5	4	27

In a study of the short-tailed shrew made in a swampy habitat near Ann Arbor, Michigan (Shull, 1907), it was estimated that the food consumption of one shrew for one month consisted of:

	Number	Percent of total
Meadow mice, or equivalent in mice	8	40
Adult insects (size of May beetles—Lachnosterna)	90	20
Insect larvae (size of May beetles—Lachnosterna)	78	20
Earthworms 4 cm. long, contracted	53	5
Snails	18	15

These proportions would undoubtedly change as any of the food species became more or less available, but it is easy to see how the fortunes of one may affect those of all the others. And the influence of such changes reached beyond the forest, for in Dr. Williams's study the shrews appeared to be the most important article in the diet of the barred owls, with mice second. Through their effect on the owls we see how the fortunes of the shrews and mice in the forest might well affect the lives of other creatures outside the forest.

The larger animals showed some interesting responses during these years of change. There were normally about 20 cottontail rabbits in the area, but in the winter of 1933 they increased to 40. The question might be raised whether this was due to the failure of other food supplies, or to some other outside pressure which drove them into the forest. These rabbits got their food from the buds and twigs of thousands of small beech and sugar-maple trees, and from the bark of many small trees. So they played their part in pruning and thinning the forest, perhaps a useful function if properly controlled. And it was soon controlled. In 1933 there were two foxes hunting in the area. The following year there were four, and the rabbit population dwindled to two. Perhaps the near disappearance of the shrews and mice played an important part in this shrinkage of rabbit numbers, for it forced the foxes to put more pressure on the rabbits.

The skunk completely disappeared following the shortage of its food supply of insects and mice in 1933, and the pilot black snake was much reduced in numbers.

So we realize that the ability of the forest to support life—its carrying capacity—is directly dependent on its ability to produce key food supplies, maintain a proper balance between the numbers of its various citizens, and give them proper shelter.

11. *life zones*

The forest community that we have seen grow up near Cleveland, Ohio, is rather typical for that area, and any forest that developed under similar natural conditions, undisturbed by man, would follow the same general course from one stage to the next. The same principles of growth and interrelationships apply to all forests. But this does not mean that all forests must have the same species of trees, for these are determined by local soil and climate. In every forest those trees survive which are best adapted to their environment.

We can see this demonstrated very clearly on a mountainside, for here both soil and climate change with the changing altitude. In some valleys it is hot, with comparatively little rain, but as we climb the slope the temperature drops and the rainfall increases. So does the amount of wind and the rate of evaporation, while the soil which holds the moisture and supports the tree roots grows thinner as we approach the higher areas, where it is newly formed from the rock.

On the different elevations of this slope we will find completely different kinds of growth, since each kind best adapted to its own special environment will survive to dominate the others. And on different sides of the mountain at the same elevation there will be variations due to prevailing winds, with their effect on rainfall, and to differing exposures to sunlight.

Thus, along the central and northern California coastline, on the west slope of the coast range, there are great stands of beautiful redwood which seem to depend on moisture from fogs that drift in from the

Pacific. On the east side of the range, where the flow of fog moisture is cut off by mountains, this tree disappears.

In the central California valleys, where the rainfall is only a few inches in a year, it is too dry for any trees to grow. Here trees give way to grasses and wild oats, for in dry country most grasses have a great advantage over trees. They have longer root systems in proportion to their leaf area and more roots near the surface, where they can absorb the slightest moisture.

On the slopes above the valley, with increasing rainfall, the growth begins to change and grasses give way to low-growing bushes or chaparral. A little higher, the growing moisture supply supports an open stand of digger pine and blue oak, growing fairly far apart at first with deep, wide root systems spreading out to find every drop of moisture in the dry soil. As the altitude increases, a heavier rainfall supports a thicker stand of trees and, among them, at 4,000 feet, the ponderosa pine appears, with its long needles and golden-brown trunks. Its numbers increase with the altitude until it displaces the digger pine. And the black oak, which sheds its leaves in winter, supplants the more delicate evergreen blue oak. Next, the sugar pine appears and increases in number.

Still higher, the shaggy Douglas fir and the pointed, compact-bodied white fir come in, soon to be joined and then displaced by the red fir. The firs are more shallow rooted than the pines, and so are fitted to grow in the shallow soil of high mountain slopes. With this reduced power to gather moisture, their lives depend on a liberal rainfall, and with the smaller root systems to support them, they must grow in thick stands for mutual protection from the wind. They are helped too by their slender, pointed tops, which shed the wind more easily than the broad-topped pines. Above them, at 8,000 to 9,000 feet, the mountain hemlock, with its flexible top, and the slender lodgepole pine take over.

At last the climate becomes too severe even for the hardiest trees. The last stragglers, their tops broken and stunted by the wind, spread out to take on the form of low bushes. Others, like the white-barked pine, bend over to grow prone against the rocks, almost like thick-trunked vines, and at last they give way to the mountain meadows of hardy grasses and low, fast-growing Alpine plants that carpet the ground with bright displays of flowers in the short summers. Above the meadows stand the rocks of the bare mountain peaks. These support primitive lichens but are not yet softened and worn by weather into a form that can support abundant life.

So we find the mountainside divided into bands or zones, each with its own type of plant community, each adapted to its own local conditions of soil and climate. And in the forest, each community is dominated by the best-adapted trees, which control and establish conditions of life for all plants, mammals, and birds that live within it.

Thus we find many different types of forest, each best adapted to the conditions of soil and climate in its own area. These conditions may be determined by changing altitude or by geographical location. The cold upper slopes of a mountain in the South may have much the same kind of climate as the low land in the North, and both will have the same types of vegetation. So, too, we find the cypress forest adapted to the southern swamp lands, while nearby, on the sandy flat lands, will grow vast pine forests where no cypress could survive.

For a single example of the adaptation of trees to their environment, compare the amounts of water used by the sahuaro, the giant tree cactus of the desert, and by the date palm, another tree that requires a very hot climate but must also have a lot of water. A 12-foot sahuaro will lose by transpiration only one fiftieth of a quart of water per day, while a date palm at a desert oasis will lose as much as 500 quarts. The palm in a single day may lose as much as 90 percent of the water that it contains within itself, while the loss by the sahuaro is only one fourth of one percent. The sahuaro lives through severe droughts on its stored reserve of water. The date palm soon dies without a steady supply.

If we watch with inquiring eyes, every tree or grass blade may tell us a story about its surroundings and its way of life.

For we now see a definite orderliness take shape out of the web of interrelationships between animal and plant, soil and climate. Vegetation in every part of the world is controlled by the climate that supplies its moisture and warmth and sunlight. And living creatures are controlled largely by vegetation. Thus, the earth's surface is divided into rather definite living communities, each one adapted to its own controlling climate. These climatically controlled areas are known to ecologists as biomes. Their boundaries may be sharply defined, as on a mountainside, or they may overlap with no sharp division; each will in many ways exert its influence on its neighboring areas, as we have seen.

We have noted the stages of succession among the trees in a forest, as each type of tree causes changes in its surroundings and so prepares the way for a new type in an orderly succession, until the permanent climax type is reached. Nearly every community of plants goes through this same orderly process of succession from pioneer to climax. Ecologists call the complete series of steps or changes in this development a sere.

in the beginning

THE MAKING OF TOP SOIL

The Grand Canyon of the Colorado River tells of the earth's changing history. The area of this canyon was once as level as the land in the background. But the rock was slowly worn away. Rain softened and dissolved its minerals. Heat and cold expanded and contracted it, causing it to crack and break into fragments which have piled up to form the grey slopes below the cliffs. (*National Park Service. Photo by George A. Grant*).

On the surface of this small glacier are rock fragments that have fallen from the cliffs above. Mixed with the moving glacier ice, these act like teeth that slowly grind away the face of the mountain.

The many layers in the face of the cliff suggest the variety of minerals that have gone into the building of this rock. These are mixed together by the ice and later spread over the land below.

During the ice ages more than seventeen million miles of the earth's surface were buried under moving ice, and under its grinding a large share of the northern soils had their beginnings. (*Glacier National Park. Photo by Hileman*).

(*Below*) Soil Building—The First Step: The grey slopes below the mountain peaks are composed of rock fragments broken from the face of the cliffs by the forces of erosion. Slowly this broken rock will be moved down into the valleys by water, wind and gravity. On the journey the pieces will be ground together and broken into particles small enough to form the parent material of soil. Before it can support life, soil must be anchored into place, enriched and organized into a living entity by the action of plants and animals. (*U. S. Forest Service*).

(*Above*) Life Makes Its Start: Lichens, moss and ferns on rock. The pioneer lichens can exist on bare rocks with very little moisture. They secrete acid which dissolves minerals from the rocks, thus creating a seed bed which can absorb moisture and provide food and a foothold for higher plants. The plant roots spread out into the surrounding soil tying it into place. From this foundation of raw soil the plant community develops. (*U. S. Forest Service*).

Plants Build the Soil: The chlorophyll in green leaves is the only substance in nature that has the power to harness the sun's energy and combine it with elements from air, water and rock into living tissue. This plant tissue becomes the food that supports all animal life, and it is the basis for the organic matter that is an essential part of productive soil. Thus, the creative work of plants is the essential first step toward building fertility into soil. By a maze of rootlets and fungus-fibres, most of them invisible to the eye, plants tie the soil together. When these die, they make a sponge of absorbent rotting vegetation. Insects, worms and other animals burrow to reach this organic food, mixing it thoroughly in the process and enriching it with their own remains. (*National Audubon Society. Photo by John O. Sumner*).

Nature Plows the Soil: Today's catastrophe prepares riches for tomorrow. In a mature forest the soil may be corrugated with ridges and hollows which hold rain and melting snow and that show where trees have fallen and mixed soil with vegetation. (*U. S. Forest Service*).

Small animals play a useful role as nature's farmers, plowing the soil. Some, like this prairie dog, dig in it to make their homes; others, like the shrews and moles, tunnel through it in their search for insects and worms, thus helping to make it rich and absorbent. (*Soil Conservation Service*).

Insects and worms do their share in soil building, plowing and mixing it with the vegetation, and enriching it with their remains as they burrow to seek shelter and food among the plant roots. This tunnel was dug by a cicada wasp to make a home for its young. (*National Audubon Society. Photo by Hal H. Harrison*).

The Finished Product—A Soil Profile: The light lower layer is the parent soil composed of rock particles. The dark upper layer is topsoil, built by the action of plants and animals. It may be as much as 40% organic matter—i.e., living and dead plant and animal life. This soil has the power to absorb water and the energy to produce crops. (*Soil Conservation Service*).

the community

A FOREST AND ITS INHABITANTS

Deer Mouse in Hole: Each creature can multiply safely only up to the limit that the forest can feed and shelter it. (*National Audubon Society. Photo by Allan D. Cruickshank*).

The food provided by plants is used to build the bodies of insects, animals and birds. Each of these in turn is eventually used as food by other living things. Thus it is passed along from mouth to mouth in a food chain to support a series of creatures, until at last it is returned to enrich the soil. Each of these young catbirds will eat its own weight in insects every day. The forest could not survive without the help of

birds. But to receive this help it must always provide enough insects. Properly controlled insects are a necessary part of a plant community. It is only when the balance between insects and controls is upset that some become a menace. (*Soil Conservation Service*).

The Sanitary Corps—Turkey Vulture with Fish: Many kinds of scavengers—birds, insects and animals—help to keep the land clean. All are necessary parts of the organization that keeps the community functioning. (*National Audubon Society. Photo by Allan D. Cruickshank*).

The ability of the land to support animal life depends directly on the food and shelter that it can provide. These tracks tell the story of a mouse searching for seeds among dead weeds. The winter landscape produces no food. It supports only those creatures that can survive on the food left over from summer. (*National Audubon Society. Photo by C. P. Fox*).

Nature automatically adjusts the number of predators to the needs of the community. These well fed deer have the strength to protect themselves and their young against the attacks of coyotes. But in the winter when the ground is covered with snow they must depend chiefly on the buds of trees and bushes for food. Only the deer that can find enough food to maintain strength can survive. (*Soil Conservation Service*).

Red shouldered hawk with snake. All animals have the power to multiply faster than their normal death rate. For its own survival, a community must have policemen to hold the multiplication of population within the limits of its food supply. (*National Audubon Society. Photo by Karl Maslowski*).

Overbrowsed Bushes: On this land there were not enough predators to protect the community. Deer multiplied beyond the capacity of the land to support them. They browsed the tree branches bare as high as they could reach. The land has lost its power to feed them next winter. The result will be starvation. (*Soil Conservation Service*).

The predator protects the community. When the deer multiply beyond the limits that the winter food supply can support, the coyote removes those that are weakened by hunger. The coyote in turn can survive only so long as there is a supply of meat to support him. Thus through the help of the predators, nature maintains a balance between each species and its food supply. (*National Audubon Society. Photo by Joe Van Wormer*).

Each creature is a specialist. This red-bellied woodpecker has feet adapted to walking on the bark of trees, a bill adapted to drilling through bark to reach the insects beneath, and tail feathers with tips fashioned to provide support while it works. (*National Audubon Society. Photo by John H. Gerard*).

The nuthatch too has feet adapted to walking on bark. Its bill is adapted to probing among bark crevices for insects, but it is not strong enough to drill through bark. (*National Audubon Society. Photo by Hal H. Harrison*).

The evening grosbeak has a big bill adapted to cracking seeds. Each of these different birds is equipped to find its own special food in its own way. Each may find its food on the same tree, but each would probably starve if it had to depend on using the food that supports any of the others. Thus each plays a special role in the community. (*National Audubon Society. Photo by Roger T. Peterson*).

Each tree is specially adapted to its own preferred environment. The saguaro cactus can store great quantities of water in its body, and exist through long periods of drought. A twelve foot saguaro can live on as little as 1/50 of a quart of water per day. A date palm, with the help of irrigation, can grow in the same desert, but it will require 500 quarts per day, twenty-five thousand times as much. (*Soil Conservation Service*).

Leaping salmon—Adapted to Excel in Its Environment; The dwellers in the streams are an integral part of the forest community. They depend on the flow of water that in turn is controlled by the conditions in the forest. The salmon spends most of its life in the ocean, growing fat on food originating with salt water plants. It has the special ability to ascend streams that are inaccessible to most fish. So at spawning time it journeys upstream to lay its eggs in the headwaters, safe from the predators of the ocean. Through this adaptation it has become one of the most successful of the fishes, threatened only by man changing its habitat. (*Hugo Schroder*).

The Struggle for Survival: This five storied nest tells of the contest between a yellow warbler and a cowbird. The cowbird lays its eggs in the nests of smaller birds which hatch and rear the young, often with fatal results to their own offspring. Here the cowbird laid its eggs in the original nest. The warbler built a roof over it for a new start, and again the cowbird laid an egg. Finally, on the fifth try, the warbler apparently gave up the struggle, and laid its own smaller eggs to be hatched with the cowbirds. Few birds are as resourceful or persistent as this warbler. (*National Audubon Society. Photo by W. V. Crich*).

The Struggle between Grass and Trees: Where trees find enough moisture they grow tall and shade out the grass. Where the moisture is insufficient grass takes it all and starves out the trees. At the borderline between grass and tree zones there is strong competition. In moist seasons the seedling trees may find a foothold in the grass, only to die in the dry seasons that follow. (*U. S. Forest Service*).

These grass roots are the key to survival in dry country. Grass has a far larger root system than trees in proportion to leaf area, and its roots are nearer the surface. This gives it an advantage over trees in dry country. Its roots take all available moisture as it seeps into the ground, allowing none to filter down to the tree roots. (*Soil Conservation Service*).

the dynamics of the community

THE RISE AND FALL OF A FOREST

(*Above*) The development of a plant community goes through a succession of different stages as each plant does its share to modify its environment, thus preparing the way for the next stage. When trees get a foothold in an open field, only those species can survive which are able to tolerate direct sunlight. Several kinds of pine, birch and poplar have this ability. Photo shows young ponderosa pine in grassy field. (*U. S. Forest Service*).

(*Right*) As the pioneer trees grow they offer shade for other species which require protection from the sunlight, such as this tulip poplar and many other species, including red oak and red maple. These trees will grow until they crowd out the sun-loving pioneer trees that nursed them. Finally their shade becomes so dense that their own seedlings die for lack of sunlight, and they in turn will be replaced by the shade-tolerant trees. (*Storer*).

Sowing the Seeds of Their Own Destruction: Trees of the intermediate stages—the oaks, tulips and others—eventually grow so big that their own seedlings cannot survive in their shade. They will be replaced by shade-tolerant trees like the beech, sugar maple and hemlock. (*U. S. Forest Service*).

The Mature Forest: The beech, sugar maple and hemlock, and other trees whose seedlings can grow in deep shade become masters. They actually select and dominate all other life in the forest. Nothing can live that is not adapted to the shade, moisture and climate that they have established. Such a forest is known as a climax forest. (*Soil Conservation Service*).

There are many types of climax communities of trees and plants, each one adapted to its own preferred conditions of soil and climate. This can be clearly seen on many mountainsides where forests are divided into bands or zones of trees, each one adapted to the conditions that prevail at its own preferred elevation. These douglas firs have small roots and great height. They have two requirements for survival: they must grow in thick stands, for protection from the wind, and they must have enough moisture to support a thick stand. Farther down the mountainside where the soil is deeper, and rainfall less, these trees will give way to a zone of ponderosa pines which have a deeper root system and require less rain. Below these again there will be other successive zones of trees, bushes and grass. (*National Park Service*).

Higher on the mountain, at timberline, the hardiest trees make their last stand against wind and cold. Spruces and white-barked pine are among the last to succumb. Above them the low, flowering plants of the mountain meadows take over, and finally these give way to the lichens that grow on the bare rocky peaks. (*Dept. of the Interior*).

Mountain forests play an important part in the world's water supply system; in many areas mountains are the source of water for all the surrounding country. Here, a current of warm, moist air blowing inland from the Pacific, strikes a California mountain range. As it flows upward over the mountain it is chilled in the cold atmosphere above. When the air is chilled it loses much of its power to hold moisture. The water vapor condenses to form a cloud, and under proper conditions, much of it will fall to the earth as rain or snow. (*National Park Service. Photo by Ansel Adams*).

Trees shade the snow, holding it in deep, slow-melting drifts. They build an absorbent covering on the ground, to hold the moisture as it becomes available, letting it out slowly to feed the streams and give life to the lowlands through summer droughts. (*National Park Service*).

Plants on the ground, under the trees, protect the soil from the hammering force of the raindrops. They tie the soil into place and make it an absorbent storage reservoir for water. (*Soil Conservation Service*).

The destruction of a forest may spell disaster to communities a thousand miles away. These trees have been blown down by a windstorm. With broken roots and weakened sap flow they can no longer withstand the attacks of bark beetles. Under a tangle of branches they are inaccessible to the woodpeckers which ordinarily catch most of the beetles that do gain entrance. (*U. S. Forest Service*).

The beetles, thus protected, multiply like an explosion, spreading out to attack and kill the healthy trees in the surrounding forest. (*U. S. Forest Service*).

The dead trees dry out and turn to tinder, becoming far more vulnerable to fire than a normal forest. The catastrophe started by a local windstorm may spread out to ruin the forests on an entire water shed. The same thing may happen when man improperly cuts or grazes a forest. (*U. S. Forest Service*).

After the Fire: This land is dead, and from it death will spread for a thousand miles over the country below, carried there by floods and droughts. (*U. S. Forest Service*).

Raindrops Spattering Soil: When raindrops strike unprotected soil, the fine soil particles spatter into the air. Falling back they fill the crevices between the larger particles, making the surface waterproof. Instead of being stored in the earth, the water now runs off in a flood, tearing away the unprotected soil as it goes. (*Soil Conservation Service*).

The Price of Removing Forest Protection: Rich top soil once lay level with the tops of these tree roots. With the coming of floods the entire farm was carried away, leaving sterile sand five feet below base of tree trunk. (*U. S. Forest Service*).

Erosion at Copper Basin of Tennessee: This man-made desert was once covered by a forest. When the forest was killed by acid from a copper smelter, the soil lost its power to store water and support life. Many attempts to replant this land have failed. (*U. S. Forest Service*).

the interdependence of communities

This man has lost his farm. It was carried away by a flood caused by the destruction of a forest on a distant watershed that he has never seen. Where has his land gone? The next picture tells the story. (*Soil Conservation Service*).

Two Farms Buried in a Single Grave: Across the center of this picture runs a dark layer of rich topsoil that once formed the fertile surface of a farm. Above it lie the layers of sterile sand and gravel washed by floods from other farms on the slopes above. The topsoil from these upper farms has been carried away, or so mixed with the gravel as to be useless. More recent floods have started to cut a new channel across the land, exposing the layers that we see here, carrying the soil down to bury other farms, or to drop it in the ocean. Total depth, 13 feet. (*Soil Conservation Service*).

A Silted Reservoir, Ruined as the Result of a Forest Fire: After a fire on the watershed above it, this reservoir, part of the water system that supplied the city of Santa Barbara, Cal., was completely filled with sediments washed off the burned hills. (*U. S. Forest Service*).

Improper plowing prepares the way for ruin. When man breaks the protecting cover of vegetation on the earth's surface, whether it be forest or grassland, he leaves the land vulnerable to destruction. Straight furrows plowed down the slope of this field form channels to carry off the rain. As it gathers speed the water tears out the topsoil. (*Soil Conservation Service*).

Dust Whirls: With its protecting humus washed away, the land becomes subject to attack by the wind. Here two whirling columns of heated air rise from the surface of a sunbaked field, lifting clouds of dust with them. When the wind starts to blow it may carry off the whole surface of the field. This can be checked by maintaining a proper mulch or cover on the surface. (*Soil Conservation Service*).

Intelligent Farming: On this farm the furrows have been drawn along the level contours of the slope, creating thousands of small reservoirs. This water will sink into the soil where it belongs, to raise new crops. And the soil will stay in place without eroding. (*Soil Conservation Service*).

Permanent Prosperity—Contoured Strips on Slope: Land properly cared for can stay productive indefinitely. It holds moisture, feeds the streams, and protects the farms and cities on lands below. The crops on the alternate strip can be rotated to allow resting and rebuilding of the soil as each crop draws its different requirements from the ground. (*Soil Conservation Service*).

12. *a strand*
of the web breaks

As one looks from a mountaintop across a vast expanse of western forest, it becomes perfectly clear that this blanket of giant trees must be the one dominant factor that controls all the life within it. But scattered through the forest blanket are occasional gray skeletons of dead trees, or patches of rusty color that tell of a dying tree. Like the animals that live beneath them, these trees seldom have a chance to die of old age, for from the time each seedling sprouts it is subject to a constant attack from animals, including insects or the smaller living organisms of disease, all seeking to harvest its store of food for their use. These attacks are a normal part of forest life.

Among the attackers is a small, dark-colored beetle, not much over a quarter inch long. It is the bark beetle, that spends much of its life under the bark of the trees. These beetles are permanent residents in many of the pine and spruce forests of the West. The growing forest withstands them through the vigor of its young trees, which can outgrow and repair the damage. Moreover, there seems to be some quality in the sap of vigorous growing trees that repels the beetles, for under normal conditions they will seldom attack a healthy young tree. As the trees become older and their growth slows, the beetles begin their attacks. Under normal conditions, most of those that gain entrance to the tree are killed by woodpeckers, which drill through the bark to get them, sometimes taking up to 95 percent. But the numbers of the woodpeckers are nicely adjusted by hunger to the numbers of the insects that support them, and if this balance is broken it may mean disaster to the whole forest.

In the White River National Forest in Colorado a heavy wind blew down several groups of Englemann spruce trees. It broke many of the tree roots, weakening the flow of sap, but left enough roots in the ground to keep the trees alive. Thousands of beetles gained entrance to these weakened trees and found a paradise to work in; for the underside of each tree, buried in a mat of crushed branches, was protected from the hunting woodpeckers. The most effective of these woodpeckers, the arctic three-toed, hunts this area chiefly in winter, but now, with the fallen trees covered by snow, the beetles were completely safe from their chief enemy and they multiplied.

First they killed the fallen trees. Then the dead trees became a focus of infection for the surrounding forest, as more beetles in search of food attacked the healthy trees nearby. The potent sap of these growing trees drove out the first attackers, but millions more followed, puncturing the bark and weakening the sap flow until later hordes gained easy entrance. These swarms outgrew the appetites of the woodpeckers, whose numbers were adjusted to a normal beetle population.

In a few years the entire forest, covering many thousands of acres, was dead or dying. Four thousand million board feet of timber stood rotting where it died, most of it wasted; for in this rough mountain country it was not worth building roads to bring it out. There was no young, productive forest left to justify the cost of these roads.

The entire forest was doomed, and all the great dependent community of living things had lost its food and shelter. This community too must go, simply because one of its smallest members had escaped its natural controls and found too much prosperity.

A forest killed by beetles will usually be replaced by another forest of the same kind, for the humus under the dead trees still offers a seedbed for new growth. But as the dead trees dry out they become as inflammable as tinder, and a bolt of lightning or a carelessly dropped cigarette may change the whole future of the area. Scattered through this dead forest in Colorado there are great areas of grassland that suggest what has sometimes happened, for these grassy areas were once covered by forest. Forest fires of the past burned the trees and destroyed the humus that had protected the land and stored moisture from rain and snow. This unprotected land faced three possibilities. On a steep slope it might have been eroded away to become a desert such as we have noted earlier. On more level land, where moisture could collect, it could have grown back to forest, following the usual slow steps of natural growth. But because the humus had been destroyed by fire, the soil had lost its capacity to absorb water, and so the grasses, which need less water than trees, were better adapted to take over.

At the borderline between forest and grassland it is often possible to see in action this conflict between trees and grass. The seedling trees spread out into the grass after a few moist seasons, only to die again when the return of drought gives the advantage to the grass, whose root system is proportionately larger and nearer the surface. Here all of the moisture from the scanty rains is captured before it can sink to the tree roots. So, the boundary line between grass and forest tends to stabilize itself, for at this point of near balance the grass, once established, can dominate the trees through its control of the moisture supply, while the trees, once established, can control the grass by their shade.

13. *the grassland community*

The grasslands, like the forests, are divided into different types, chiefly dependent on the quality of the soil and the amount of moisture they receive. Most of the country's original stands of grass have been changed by plowing or grazing, so we cannot see them as they originally stood. But even today there is enough native grass left to tell its own story.

Since the high mountains of the West cut off much of the moisture that drifts in from the Pacific, the Far Western plains receive little rainfall. To the east the rainfall increases, as warm air moving up from the Gulf of Mexico brings more moisture, and with this added supply the country supports taller grasses. Thus the grass of the original prairies was divided into three main belts.

In the central part of the continent a belt of tall grass prairie ran from Texas to Manitoba, with stalks ranging from five to ten feet in height—big and little bluestem, Indian grass, and others, forming a heavy sod that built through the ages some of the world's most fertile soil. West of this area, adapted to the decreasing rainfall, lay a great belt of mixed grass prairie. It carried shorter grass running up to four feet in height—little bluestem, June grass, wheat grass, and several others. Most of these did not form sod but grew in bunches, which gave space between for a great variety of flowering herbs. These included legumes that carried on their roots nitrogen-fixing bacteria which helped to enrich the soil.

Still farther west, this mixed grass belt gave way to the "short grass" prairies of the high plains. Here, screened by the mountains, there was so little rainfall that the moisture in the soil seldom penetrated below twelve or fifteen inches. It evaporated or was used by the grass before it could sink any farther. So in this dry region the minerals dissolved from the soil particles were not leached away by the water, and the earth holds a rich accumulation of plant food. This mineral-rich soil produces some of the world's best forage for raising meat. Covered by drought-resistant grass, developed through the ages, it is natural grazing land. But under our farming system this rich dry land carries the seeds of its own destruction, for in moist seasons these fertile soils can produce big crops of wheat. This tempts many farmers to plow the land, making it vulnerable to disaster when a period of drought and crop failure leaves the broken surface unprotected by any covering. For now the wind can carry away the topsoil in great dust storms, leaving behind it a potential desert. Thus the country's dust bowls have developed.

We have seen that the boundary line between grass and forest is fairly stable. But not so with the line between grass and dust bowl. Here there is fierce competition in the dry seasons as the rolling dust spreads out to engulf the grassland in its path, while in wet seasons the soil may be moistened long enough to start a new stand of protecting grass.

We have seen that conditions established by the trees or by the grass dominate the lives of all the creatures that live with them, both through the kinds of food they manufacture, and the shelter that they offer. And so we find different types of creatures in the forest and in the grassland, each best adapted to its own environment.

Since the tree is composed largely of wood, which

is, after all, chiefly air, water and sunlight, organized by chlorophyll into solid form, it needs little mineral food, and it is well adapted to growing in areas of heavy rainfall, where the surplus water leaches the minerals away. But trees, or other plants, grown on such soil are poor in the mineral foods that animals need to support life. So the food offered by the trees is fitted only for certain kinds of animals. First, the leaves, most of them inaccessible to larger animals, are eaten chiefly by the leaf-eating insects; second, the wood and bark are eaten by boring insects, and third, there are the buds, seeds, and fruits, which contain the stored supplies of nutrients to support new growth. These are eaten chiefly by birds and by small animals like squirrels and mice, for it takes a small and active animal to gather them from the branches and to find enough of them to support life. There are few large animals in the forest.

Creatures that eat plants are known as key-industry animals, for they reorganize nature's basic plant food prepared by the leaves into richer, more concentrated food that will support flesh-eating creatures. Like the squirrels, grouse and deer are also key-industry woodland creatures, but the deer is too large to find enough food in the deep forest. It lives chiefly around the forest borders, where it finds some grass and a rich growth of bushes to supply the buds and leaves it needs for its main diet.

The animals of the grasslands are adapted to a very different diet, for the grasses and range plants, growing on soil far richer in available minerals, prepare a food that is better suited to support animal life, and they offer it in quantities easily harvested to support big bodies. So we might think of the forest animals as machines adapted to very active work in the search for small amounts of food, and we might think of the range animals, like the buffalo and antelope, as machines specialized for the reorganization of large amounts of rich plant food into the building of large and active bodies that can protect their stores of meat from flesh eaters. For the animals of the grasslands lack the shelter of the forest. Their safety depends on strength or fleetness of foot or, in the case of the smaller animals, on ability to find shelter in the ground.

This community of grassland animals is organized in its own special way to meet its own conditions. But it is subject to the same basic principles of adaptation and balanced control that apply to all living communities.

When the white man first came to the prairies he found the buffalo as the chief grassland animal. The antelope took grass, but lived chiefly on sagebrush on the drier lands, while the elk lived mainly near the forest, for it needed some buds and leaves from the trees to supplement its main diet of grass.

As these heavy animals in great herds trampled the land, they changed the character of the soil, compressing it and filling the porous spaces left by decaying grass roots. This both lowered its ability to absorb water and shut out the air required by grass roots for growth. In time this loss of air and moisture would have seriously affected the crop of grass, which would gradually have been replaced by tough-rooted weeds, less nutritious and able to support fewer animals.

But, fortunately for this community, there were other smaller animals living in the soil. Some of these, like the gophers, lived on the roots of plants; others, like the ground squirrels, lived chiefly on leaves, while still others, such as the kangaroo rat, specialized in seeds. All these creatures lived in holes in the ground, and in the course of their digging gave the earth a very thorough loosening and turning over. In some places, where this plowing action has been measured, it was found that the ground rodents gave the soil the equivalent of a complete plowing in as little as twenty years. This slow process allowed the grass to maintain a good growth on the surface, while still getting the benefits of the plowing. Hence these small animals are really nature's soil builders in dry country, just as truly as are the worms, insects, shrews, and moles of moister land.

This useful work required great armies of rodents, and in bad years, when dry weather reduced the grass crop, they starved by the million. So, one requirement for their survival was that they should have the power of multiplying very fast to replace their losses. This multiplication was usually controlled by disease when the land became too crowded. But like the creatures in the forest, these animals too were fairly well controlled by predators. Snakes, coyotes, owls, and many kinds of hawks all played a very important part.

In the year 1910 I saw a field of alfalfa being harvested in Montana. There was a fence around the field, and on every fence post there sat a hawk that swooped down from time to time to pick up the meadow mice as they scurried for shelter from the cut hay. There were discarded pieces of mice scattered around the base of each post, and hawks patrolled behind the mower to pick up the mice exposed by the cutter bar.

That was over forty years ago; but since then the hawks have been pretty thoroughly extirpated over large areas by sportsmen and farmers who didn't recognize their best friends when they saw them.

The larger predators, the coyotes and wolves, also played a very important role, for the grass required protection from the large grazing animals too, and these predators helped to control their numbers. The grass grows its annual harvest of food, which can be eaten by the animals, but it is also the factory that produces the harvest. To maintain this factory in good condition, enough must always be left standing after grazing to build new leaves and support the roots in good health. The factory must never be harvested with the crop; in nature's organization, predators perform the function of protecting the factory from grazing animals. When undisturbed by man or by some catastrophe, this protective system adapts itself to maintain a fairly good balance.

In a study made at Yellowstone Park, Adolph Murie found that on a good range the well-fed deer can protect themselves and their young from attacks by coyotes. During the summer, therefore, coyotes live chiefly on smaller creatures. But in a severe winter, when snow covers their forage, the deer that are weakened by hunger can be easily killed. And now the coyotes reduce their numbers to those that the winter range can support in good health. So, when coyotes kill deer in winter it is a very good indication that there are more deer than the range can support.

And the coyotes' numbers are controlled by hunger, too, for there are two basic laws that regulate the numbers of all predatory animals. First, in order to survive, the animal must always find enough food to carry it to the climax of its next hunt. Second, in order to multiply, it must find enough added food to rear its young.

So, in a mild winter, when the deer find enough food and strength to protect themselves, some coyotes may die for lack of meat or, in their weakened condition, succumb more easily to parasites and disease. And, of those that survive, some will fail to rear young. In this way, protected by predators, the community of plants and animals maintains its balance. Before the coming of the white man, predators kept prairie herds down to the numbers that the winter range could support. The properly controlled herds were well fed by the luxuriant growth of the rich summer grass, whose tasty upper leaves they harvested; then they moved on to fresh pasture, leaving the tough, unappetizing lower leaves of the factory to produce another crop.

So each member in the grassland community, as in the forest, plays its special useful role, and the value of each member depends on the action of other members which interact upon each other to keep the whole in balance.

Many of man's problems today have been brought about by ignorance of this fact. For example, there is the famous record of the Kaibab Forest in northern Arizona, covering a little over 700,000 acres. In 1905 the stock of deer was estimated to be about 4,000, while the potential carrying capacity of the range was nearer 30,000. In an effort to increase the number of deer, the coyotes, wolves, and mountain lions were killed off. Between 1907 and 1917, 600 lions were removed and in the next few years 116 more were killed. Beginning in 1907, 11 wolves were killed, and between 1907 and 1939 more than 7,000 coyotes. With this removal of predators, the deer began to increase. By 1918 the population had gone up to over 40,000 and the food plants on the range began to show signs of damage. By 1920 some of the fawns were starving. By 1923 the herd had increased to an estimated 100,000, far above the normal carrying capacity of the range. The forage on the range was very much damaged, lessening still more its carrying capacity, and during the next two winters about 60,000 deer starved to death. The reduction continued until by 1939 the herd had dropped to about 10,000, struggling to survive on a very much damaged range. While there were, perhaps, too many of the predators at the beginning of this experiment and the deer were below the range capacity, it may well have been that part of this reduction was due to overhunting by man. It is obvious that the killing of the predators resulted in the ruin of the range, reducing its power to support deer. Actually, predators in proper numbers had protected both the range and the deer.

14. *some laws of life*

In the development of plant and animal communities a few guiding principles stand out, which we might consider as laws of life.

The first of these is *the law of adaptation*. In every area that can support life, communities of plants and animals develop which are best adapted to this particular environment. They live here, not necessarily because they are the only ones that could survive here, but because they are able to excel all competitors in the struggle for available food and shelter.

But sometimes, like the trees that we have watched in the forest, these best-adapted forms will sow the seeds of their own destruction, for they bring about changes in their environment until it becomes better adapted to support other forms, and finally they will be displaced by these new forms. So *the law of succession* plays its part too.

We have seen that every living thing is subject to so many hazards that its race might be wiped out by some catastrophe if there were no insurance for its recovery. But living things are given this insurance in the form of great powers of multiplication, which can make up for great losses. We might think of this as *the law of multiplication*.

We have seen too, that even the smallest of living things, like bacteria, could overrun the world if their multiplication were unchecked. But in every case the world is saved from this catastrophe by some form of control. The safety valve may be in the form of disease, predators, or just lack of food or shelter. So there is *the law of control*.

Under these laws the community eventually approaches a sort of balance where the forces of adaptation, multiplication, succession, and control work together to produce on the land the greatest amount of life that it can support, which might be called the measure of its greatest carrying capacity.

Sometimes this carrying capacity may be very greatly affected by special conditions in the environment, called limiting factors.

For example, a study was made of four lakes in Wisconsin. Two of them had soft water, containing very few minerals in solution. This lack was a definite limiting factor in the carrying capacity of the lakes. The other two had hard water, containing mineral solutions drawn from the surrounding soil and washed off nearby land. The minerals in these hard-water lakes changed the entire life of the community, for they acted as fertilizer to produce from three to five times as much plant life as the soft water. The animal life, exclusive of fish, was two to three times greater, partly because the plants gave more food to support the animals and partly because the water supplied more minerals to build their bodies, including shells for the molluscs. The remains of dead creatures also fertilized the water, to produce even bigger crops of plants. And so, in these fertile lakes and on the land around them, one unseen step led to another to build a richer life.

Studies were made at Lake Erie to determine the reasons for the great annual variations in the numbers of certain kinds of fish, including the white bass, the yellow pickerel, and the cisco. It was found that these variations coincided very closely with the fluctuations in the quantity of plankton—the microscopic plants that build food in the water, and the microscopic animals that feed on them.

Since the cisco live entirely on plankton, the interdependence here was plain. But the bass and pickerel through most of their lives live on fish and other creatures larger than plankton. Why, then, should the plankton affect their welfare? It was found that there is a stage in the life of the young bass and pickerel, soon after they are hatched, when they, too, depend almost entirely on plankton, and the supply of this microscopic food during these few critical weeks was the limiting factor which controlled their numbers. Plankton also feeds the small fish which later supply the larger ones with food. Actually it is the foundation stone of the fish supply in the lake.

And what is the cause of this variation in the supply of plankton? It was found that the abundance of it varied in direct relationship to the amount of rainfall on the land. A heavy rainfall carried into the lake large supplies of mineral matter, including lime, phosphorus, and potash dissolved from the soil, supplying new food for plankton. Under natural conditions, then, as we have seen, when rain falls on healthy land, it contributes to the benefit of the entire fish population of the lake.

But plankton cannot live without sunlight, which gives it the energy to convert the elements into food; and sunlight can reach the plankton only if the water is reasonably clear. With overgrazing and improper management of farm lands on watersheds that supply the lake, heavy rains now erode the land and fill the water with mud and silt, which cut off life-giving sunlight.

So, improper farming on the watersheds has changed life-supporting rain into an agency that destroys the fish and the livelihood of the many fishermen that depended on them. This improper farming is a limiting factor in the productivity of the lake.

In the Okefenokee Swamp in Georgia we find some dramatic examples of plant successions, and of the influence of life on its environment. Here, in waist-deep water there grow great gardens of water lilies and the neverwets, aquatic relatives of the calla lily. On the swamp floor beneath them is a deep deposit of muck, laid down each year, through the ages, by the dead ancestors of these living plants.

As this deposit builds up toward the surface, plants of the shallower water find a foothold on it to crowd out the lilies and the neverwets, so we find toward the shore a growth of shallow-water plants such as pickerel weed, maiden-cane grass, and button-bush. At the water's edge the pickerel weed and others give way to the giant pitcher plant, which must have firm footing for its roots yet still keep them in the water. Behind the pitcher plants, in the shade of the trees, lie great beds of ferns. And here the cypress trees are making a losing fight for survival. They started life in the water, but as the older ones died and gave their substance to fill the swamp, they slowly built dry land above the water and prepared the way for the destruction of their kind, for at length trees better adapted to dry land, such as the bay tree, will come in and crowd the cypress out. We can see examples of plant successions like this in almost any marsh or pond that has been undisturbed by man for a few years.

Water animals, too, can play a dramatic part in the building of the landscape. It has been estimated that about one-tenth of all the flat meadowland in the Rocky Mountain area has been built by the silt and organic remains that collected through the years in the ponds made by beavers. These beavers have caused far-reaching changes in the land itself, transforming stream valleys into broad and fertile meadows that favor the development of an entirely different kind of community life. For this meadowland now contains the rich organic remains of plants that grew in the quiet waters of the old beaver ponds. The beavers have carried the land forward a long step in the building of fertile soil, thus preparing the way for the next step in the succession. In this case we can see the broad outlines of the change, but the life-supporting quality of land or water depends on many things that the eye cannot see.

The state of Florida is one of the great cattle-growing areas of North America. With its vast stretches of fertile, well-watered soil and its mild winters, it would seem ideally suited for raising cattle. Yet, for many years all attempts to raise them in Florida were failures. The thoroughbred cattle of the western plains could not survive conditions on the Florida prairies. The scrawny native stock, descended from generations of tough ancestors, slowly starved on Florida vegetation.

Then, around the year 1940, a series of tests showed that if the cattle were fed very small amounts of cobalt, a fraction of an ounce to a ton of hay, they would survive and do fairly well. Cattle do not need cobalt in their own make-up but, as with many other animals, their digestive processes depend on the help of microscopic organisms living within the body. And it was found that in cattle and sheep and some other animals, these tiny aids to digestion need cobalt to carry on their functions.

This discovery, while it helped, did not really solve the problem for the stockmen. The cattle no longer starved, but they still were not at their best in Florida, with its humid climate and its hungry insects. However, in parts of far-away India the climate was not unlike that of Florida, and Brahman cattle there had developed the ability to withstand heat and humidity. Experiments showed that these cattle could do well in Florida, especially when new grasses were planted to improve the range.

So at last the stockmen came into their own, with a synthetic environment and an imported cow adapted to live in it. And with this new combination a great industry was opened up to the state of Florida.

There are many factors like this, which we would never suspect, that may control the life of a community.

For example, how many bobwhite quail can a certain brushy field support? It produces enough food and shelter to support about a dozen birds. Two years ago it carried a family of twelve. Its population had been kept to about that safe limit for years with the help of an occasional visiting fox or hawk. In the winter its members had warmed each other by sitting in a circle at night, each bird protecting its two neighbors from the wind and sharing its warmth. An occasional bird nestling on top of the group gave extra protection. When a predator attacked the huddling birds, they burst outward like an explosion, so confusing the attacker that they were usually out of reach before it had singled out a victim.

But one fall a hunter found the covey, and bagged eight birds. The four survivors were numbed by the cold in a winter storm, and one of the dazed birds was caught by a fox. A little later the other three froze to death.

These birds had been safe only so long as there were enough for mutual protection. Their ability to exist in this field had been limited by two factors—they could not survive at all unless there were enough of them to give the needed mutual protection, and their numbers must still be kept within the limits that the field could shelter and support.

The musk ox on the northern tundra, in their normal numbers, were once fairly safe from attacks by wolves, for when they were attacked they would gather in groups, facing out, with the young in the center, and in this formation they could protect one another.

But when human hunters reduced them to bands too small for self-defense, they became easy prey to groups of wolves, which would surround individuals and attack from behind.

Each such artificial reduction in musk ox numbers, without a corresponding reduction of the wolves, meant a reduction in the normal food supply of the wolves, and therefore an added pressure by the hungry animals on the remaining musk ox.

Certain species of termites depend entirely on cellulose as raw material for their food. They have not the power to digest this cellulose by themselves, but they carry in their intestines special kinds of protozoa, microscopic single-celled animals, which digest the food and so make it available for their use. The protozoa cannot live in a temperature above 36°C. The termites themselves can stand higher temperatures but they will soon starve to death if their life-giving partners are killed by heat. However, starvation can be prevented by a new supply of the proper protozoa, which will quickly restore them to normal health. So the termites supply food and shelter for their internal guests and these, in turn, make life possible for their hosts, so long as they are given a cool enough environment.

This relationship of the termites and their bacteria may seem especially intimate, but it only helps to emphasize the very close interdependence between each member of the community and its neighbors and the rest of the environment. The laws that govern this community interdependence might be summed up like this: first, the *limiting factors*, which determine what forms of life the environment may or may not support; next, the *adaptation*, which determines how these forms may fit into the environment. The adaptation may be limited by specialization, which leads to *succession* of species as the environment develops and changes. The survival of the species under adversity is provided for by high *multiplication* rates, and the dangers of multiplication are guarded against by *controls*. The carrying capacity or productive capacity of the environment depends upon the proper balance in the working out of all these basic laws.

15. *arteries of life*

Of all the many things that go into the making of life, water is the one most prodigally used. It not only makes up a big proportion of all living substance, but it is steadily used up to maintain the life process and must be as constantly renewed. The liquid that an animal drinks is only a small fraction of the water that goes into the maintenance of its life.

It is hard to make an accurate measurement of all the water used. But, as we have seen earlier, it is estimated for example that it takes about 900 pounds of water to produce a single pound of dried alfalfa, and about 5,000 pounds to produce one pound of wheat. Translate the plant into meat, and it takes about 5,000 pounds of water to produce one pound of beef.

The supply of water directly controls the kinds of plants and animals that can live in any environment, and so in any community the source of its water supply is a most fundamental part of the environment.

Many parts of the earth's surface receive but little rain. In some places the loss of moisture through evaporation is greater than the entire rainfall. So, existence here is totally dependent on the arteries that bring this life blood from the community's heart in the distant mountains and distribute it to actively functioning members in the dry lowlands.

When we speak of nature's arteries, it would be natural to think of rivers. Actually, as far as dry-land communities are concerned, by the time water has reached a river its value is largely gone. The river is carrying the water back to the ocean just as fast as it can get it there. To be useful on dry land, water must be delayed on its journey to the sea, and distributed so as to render the greatest amount of service.

When rain strikes a bare slope its natural course is to go straight downhill by the shortest way, and into the nearest river. It may leave a little moisture along the way, but the land is soon as dry as if it had never rained. There is no water left to support life, and except for primitive forms, there is no life to use water when it does rain. We have seen how such a condition developed on the man-made desert in Tennessee. We have seen, too, how nature goes about building a protective sponge on the earth's surface to control run-off. We will now see how the life of a community in the lowland depends on the effective working of this sponge in the higher country that supplies the water.

Our western mountains receive a good part of their water in the form of snow. High in the mountains of Colorado, at the headwaters of the Colorado River, a snowflake coming to rest on a spruce branch may lie for days where it falls, buried among its neighboring flakes. As it melts, some of its moisture evaporates back into the air, for water evaporates fast in the thin, decompressed air of the high mountains. The rest of the melting flake eventually flows down the tree trunk, or drips off onto the snow below. Here, shaded from the sun, the melting of the banked snow may take weeks or months. The water reaches the soil slowly. An insulating blanket of spruce needles protects the soil from freezing. Thus its open pores are always ready to receive water, which spreads out to fill every crevice between the soil particles and to filter into the humus, which holds it like a sponge.

As it fills the soil, part of this moisture adheres to the surfaces of every particle of mineral matter and humus, while the rest continues to sink slowly, following the pull of gravity.

Thus it is that at last the hillside becomes filled with a huge store of water, part of it anchored in place, the balance moving slowly downward, a great, leisurely, underground river, as the unhurried melting of the snow sends in fresh supplies from above to maintain its flow.

The capacity of the ground to hold this vast moving reservoir depends chiefly on the quality and depth of the soil, for a good soil can hold the equivalent of more than a third of its bulk in water.

A good share of this stored water will go into maintaining the organization: nourishing the vegetation that makes the reservoir possible, building and protecting the soil, helping to grow timber and grass.

The remainder continues on its way downward, into the flat dry lands below the mountains. Here there is little rain or snow to moisten the surface, but moisture from the moving water in the ground rises by capillary action to supply thirsty vegetation, and where the water sinks too deep for this, longer-rooted plants reach far down for it.

As the earth in the valley fills with water, some of it may break out on the surface in springs to join the brooks that carry the overflow from the mountains. Many of these springs rise unseen, to flow into the stream beneath the surface. Thus slowly the stream swells as it travels, until at last it joins the Colorado River, a boiling turmoil of flowing silt mixed with water, "too thin to plow, too thick to drink," carrying the run-off from nearly a quarter of a million square miles of land.

This mixture of silt and water might tell us an interesting story about the land it comes from. It carries to its destination in Lake Mead, above Hoover Dam, an average of nearly seven hundred thousand tons of silt each day throughout the year. All of this great mountain of dirt has been scoured off the hillsides, covering thousands of miles of watershed, and each ton represents a bit of land that might have served as a reservoir to store the water and control its flow.

The flow of the river may rise to more than sixty million gallons a minute in summer when the snow is melting on the mountains, and may sink to about two million gallons in the winter when the surplus water has run out. This flow has always varied with the seasons, and it has always carried great quantities of silt. But the load of silt has greatly increased in recent years with destruction by fire and overgrazing of nature's protective cover on the hillsides. And with the increasing erosion of the natural storage reservoir, the extremes of flood and drought have greatly increased. So the life-giving potentialities of the river in the lowlands are greatly affected by conditions on the watersheds, many hundreds of miles away.

16. *land and water meet:*
the pacific salmon

Of all the world's gardens the largest
and most productive are to be found growing in
water. Most of the plants that fill these liquid gardens
are the microscopic plankton, too small to be seen
by the naked eye. But they grow in incredible numbers, so that their masses darken the water in many
parts of the ocean. On this food all the creatures of
the sea depend for their existence.

Some, like the sardines, eat plants, transforming
them into the meat and bone of their bodies. These
in turn become the food that supports carnivorous
fishes. Taken all together, the living creatures of the
water supply about one third of all the food that supports mankind.

In the ocean's vast maelstrom of living creatures
no life is safe from attack. The eggs or young of the
largest may offer food to the smallest. Those that survive to reach maturity form only a tiny fraction of
the total.

Through the centuries many forms of fish have
developed habits of life that help them to prosper in
this fierce struggle for existence. Among the most
successful are the five species of salmon of the Pacific
Ocean which use the watersheds of western North
America for their breeding grounds.

These salmon spend most of their lives in the
ocean, growing fat on plankton and the living things
that it supports. Not very much is known about their

lives in the salt water, but experiments with marked fish have shown that some of them travel long distances. Many of the chinook salmon hatched in the Columbia River are caught more than a thousand miles away, off the coast of Alaska.

When spawning time approaches, each fish returns to the river where it was hatched. And here the story of its life becomes an open book. For now the salmon leave the salt water and journey up the streams of the North Pacific coast, to find a safe nursery for their eggs.

In these clear, cold waters flowing from the glaciers and snowbanks of the mountains, there is little food to support the predators that might otherwise survive in fresh water, and none but the hardiest can surmount the rapids and falls that block the passage to the spawning grounds. But nature has provided the salmon with the ability to overcome all these obstacles.

As spawning time approaches, the salmon congregate along the coast and swarm into the streams. In some places they crowd so close that the water appears to be a solid mass of fish, and it seems that their numbers are limited only to the capacity of the water to hold their bodies. Here, one of nature's miracles begins. This swiftly flowing water holds little food. It could not possibly supply enough for all these fish. But, as they reach fresh water, a change comes over the salmon: they lose the power or the desire to eat. But their bodies are well supplied with fat that has been stored up during their time at sea, and on this stored energy they continue their journey onward.

But how can this enormous mass of fish find room to spawn in such a narrow channel? Must each one fight for its place until the strongest drive the others out? That would be a terrific waste of the stored energy that must now support them for the rest of their lives.

The problem is solved by another of nature's miracles. Among the salmon that enter the streams there are five different species: the great chinook (or king), largest of the salmon, sometimes reaching a weight of a hundred pounds. Smaller than these are the silver (or coho), the chum (or dog), the sockeye (or red), and the hump (or pink)—each species with its own special characteristics and its own preferences as to the time and place of spawning. Studies of these fish have shown that among each species there are many different races which we might think of as tribal groups, each with an inherited instinct to return to its own parent stream and, within that stream, to the particular area where it was hatched.

Thus, out of this milling mass of fish, order will soon begin to appear. The first to enter are usually not attracted by the nearer, more accessible spawning grounds. Instinct has brought them here early because they have a longer journey ahead of them. They move onward, up through foaming cascades, leaping barriers of rock, fighting their way through rapids, sometimes packed so close that they give the appearance of a solid moving escalator built of fish. Here some may leap out from the struggling mass and flap their way over the backs of their fellow travelers till they find an opening into the water again.

Along the way many find themselves caught among the rocks, where they gasp and die. Others furnish food for predators of many kinds—the wolf, the bear, the otter.

In quiet areas they pause to rest, and at last they settle for their spawning as they reach the ancestral breeding grounds, marked, perhaps, by the temperature of icy water flowing from a snowbank, or perhaps by warmer water from a lake flavored by chemicals from decaying vegetation. The velocity of the current and the character of the bottom are important factors in their choice of a site for the nest, or redd. For the current must be swift enough to sweep the bottom clear of sand, leaving good-sized pebbles exposed, with just enough sand and gravel in the crevices to serve the needs of redd building.

Here a pair of chum or dog salmon are preparing for their spawning. The lady does the work, fanning the bottom with her tail to dust out the gravel from among the larger pebbles. This is carried away by the current. As the work progresses, she lies on her side, head and tail against the bottom, then with a vigorous lift of her midsection she creates a suction along the bottom which lifts the gravel from the deeper crevices among the pebbles. As the bottom becomes clearer she feels over it with her tail and lower fins to learn when it is ready for the spawn.

This work may take several days, and during its progress the head of the family watches over it, driving away the intruders that would take his place and being very attentive to his partner. He brushes against her side, swimming back and forth over her and at times settling by her side to quiver until, in shallow water, the drops of water fly from his exposed, vibrating back. When the redd is nearly finished the interest of the male becomes more intense. Two or three competitors may try to crowd out the guardian male, who drives them off with jaw or body or tail. If they become too importunate the lady helps him, apparently recognizing his established right as her protector.

At last, when the nest is ready, the female settles on the bottom, feeling with her fins for a suitable crevice between the pebbles to receive the spawn. Now a male settles on each side of her, and for the moment she accepts them both. As the eggs are dropped into the crevice, the water turns white with the milt from the males that fertilizes the eggs. Then comes a burst of feverish activity as the female dusts gravel from the bottom just ahead of the nest. This drifts down to fill the crevices and bury the eggs to protect them from hungry trout or mergansers that would eat them.

In this laying there are a few dozen eggs. As the female dusts fresh gravel over them, she uncovers a new patch of pebbles, which will now be prepared to receive more eggs. In like manner a series of separate egg deposits is made, until all are laid. This may total about 1,500 eggs in the case of the hump salmon, and up to several thousand in the case of the big chinooks.

During the course of this activity the fish slowly waste away. The back fins disintegrate until nothing is left but a row of bones standing above the back. Patches of skin turn gray and drop off. Finally the whole body turns white and there remains barely enough energy to continue work. At last both male and female turn belly up or sink to the bottom to drift with the current, struggling feebly until they land among the windrows of their fellows that pile up along the banks.

Among the Pacific salmon no fish ever survives to make the return journey to the sea. No matter whether the journey is ten miles up a small tributary or two thousand up the Yukon, each fish comes to the river with just the necessary store of energy to carry it to its inherited destination and to finish its work. There is no strength left to carry it back to the ocean.

But the bed of the river, with its tributaries, is filled in every suitable spot with uncounted millions of eggs buried under four to six inches of pebbles, safe from the roaring floods and rolling rocks.

Now, through the winter a new drama unfolds, hidden in the gravel beds under the water. This is the beginning of a new life inside the egg. The egg is transparent, and in a hatchery the whole process can be followed by an observer.

First the eyes appear and the blood vessels that supply them with nourishment. Slowly the dim outline of a baby fish takes shape. Near the throat a small red dot appears and begins to pulsate. At last the egg membrane breaks and the tiny fish emerges, still carrying the egg yolk that will supply its food for two or three months under the protecting gravel.

The body grows, building fins and bones and internal organs out of the material supplied by the yolk. Finally the baby fish struggles up through the gravel of the stream bed to find its own way in life and to gather its own food, starting with the microscopic plankton in the water.

But the mountain streams supply little food and cannot support even these tiny fish in great numbers. The number that a stream can carry will vary considerably. Water flowing from regions of insoluble igneous rock will have very little dissolved mineral matter to offer the plankton for fertilizer, and such streams can supply little food to support fish. In other areas, rocks may offer plenty of lime or phosphates to fertilize the water, and here the streams will produce more food and more fish.

Nature adjusts the number of fish to the available food supply in various ways. The young hump salmon and dog salmon, which spawn fairly near the ocean, return to the salt water almost as soon as they leave the redd, while other species remain in the river. The sockeye salmon spawns only in streams that have lakes along their course, and their young spend a year or more in a lake, where food is more plentiful, before heading for the ocean.

The young chinook and silver salmon spend the first year or two in their parent stream, and it is possible to make a fairly full study of the early lives of these fish.

It has been found that, on the average, out of each 5,000 eggs laid by a female chinook, about 4,000 will hatch. From those that hatch, only one in ten will live to reach the ocean. The rest die—they are eaten by trout, merganser, or other predator, or starve, or are trapped in a drying pool. Out of the 400 that reach salt water, only one in every hundred will escape the hazards of the ocean. So, in the end, out of each 5,000 eggs laid by the female chinook and fertilized by her partner, only four fish will survive to return to the parent stream for spawning. But the others too will have played their necessary parts in the web of life, giving their substance to feed the myriads of other creatures of sea and air and land, contributing as do all the rest to support the great organized pattern of life on earth.

About six hundred million pounds of salmon are caught every year on the North Pacific coast. A good part of this production depends on the character of the watersheds from which the salmon streams receive their water supply. As we have seen, rocks provide important minerals and the quality of the water is influenced by many other factors which are determined by the nature of the land.

When a salmon builds its nest on a gravel bar, it has no way of knowing what the future may hold in store for that gravel bar, and it does not care. But to the future of the eggs this is a most important question. If the supply of water becomes too low, and the gravel bar is exposed to the air, the eggs may freeze or dry out, or in very high water the bar may be washed away. If the water level falls after the young hatch, thousands may die in the drying pools. If these catastrophes happen too often on a river, there will be no young salmon to carry on the stock, and no salmon run. The inherited instinct to return to the ancestral stream has been a key to the survival of the salmon up those rivers that have been able to raise their ancestors. But many catastrophes may happen to the watershed that produces a river, and may destroy its ability to maintain the salmon run.

The broken debris from a cut-over forest may collect to build a log jam that can block the salmon and wipe out the races that once spawned there. The burning of a forest will destroy the ability of the land to store water and regulate the flow of the river. Resulting floods and low water may destroy the spawning beds or kill the young fish. Land that is exposed to erosion after a fire or improper farming may deposit so much silt in the stream bed as to suffocate the young fish. This has been considered one of the most important causes for the disappearance of the Atlantic salmon from the rivers of New England.

The Ariel dam, on the Lewis River in the state of Washington, was built not far above a salmon hatchery. After the reservoir was filled above the dam, there was a 100 percent loss of young salmon and eggs in the hatchery. This was caused by the decaying vegetation in the reservoir, which changed the chemistry of the water. The hatchery had to be abandoned until, after several years, the decay was completed. The cause of the disaster was apparent, and it suggests to us what a tremendous unseen loss might occur in a river as a result of very minor changes in the quality of the water. One of the chief purposes of a dam is to store water and let it out when needed. The alternate discharge and holding of the water may cause great changes in the water level and so kill millions of eggs and young fish if not properly controlled.

If a dam is not too high, and is provided with fish ladders, the salmon can pass over it to spawn above. But the climbing of each dam takes its toll in time and strength, and there is some evidence that if there are several dams on a river, the fish will not have the time and strength to reach their normal spawning grounds, and will die before laying their eggs. If the parent fish do succeed in spawning, the dam takes its toll of the young fish on their journey to the sea. No sure method has been devised to take them back over the dam, and the tremendous pressure changes in going through the dam discharge kill a great many. Attempts to measure the mortality indicate that about 15 percent of all the young fish trying to pass a high dam will be killed. This may not seem a very big loss, but it does not take very many dams at this rate to kill most of the fish.

One of the greatest hazards to young fish is the pollution of water by poisonous discharges from industrial plants and from city sewage. Most of these dangers can be eliminated by proper disposal of the wastes.

Another great cause of loss lies in unscreened and improperly used irrigation ditches. Millions of young fish get lost in these on their journey to the ocean, to die in the fields. Millions more are killed in the parent stream when water is allowed to run off into the fields unnecessarily.

The irrigation ditches can be screened to keep out the young fish, but proper screening is expensive and in many places it is neglected. In regions where there is plenty of water for irrigation it is easy to let the surplus water continue flowing down the ditches, thus lowering the stream flow.

In this way, as in many others, the careless use of a natural resource by one man may ruin the livelihood of another.

The story of the salmon suggests some interesting thoughts on the use of our natural resources. Here, one of the world's great sources of food was largely destroyed in the United States, but is now receiving intensive study; valiant efforts are being made to revive it. Pollution of the rivers has killed millions of fish. The pollution is being reduced under compulsion of the law. In the building of dams there is sometimes a choice of available sites. One site may be on a branch of a river where few fish spawn, another may cut off a valuable spawning area. The choice of site may involve conflicting interests—of farmers for irrigation, of cities for power, of forest lands for access roads. Attempts are being made on the Columbia River to consult and study all the interests involved in working out the long-range plans for dam building.

In Alaska, intelligent efforts are being made by the canning industry to achieve the highest possible sustained production from salmon resources. A research organization has been set up at the University of Washington, financed by the canneries, to study all the different factors in the life of the salmon, to learn how much can be produced by a given stream, and how the stream can be managed to give the greatest yield. The harvesting of the salmon crop is carefully regulated, so as to insure that each race entering the stream shall be allowed to reproduce enough to insure a good crop for the future, and that only the surplus fish shall be caught. This is a step toward a really efficient, scientific use of a major natural resource.

17. *the evolution of life*

We have seen how all life faces a constant struggle for survival which, in the long run, is won by those individuals and species that are best adapted to the conditions of their environment. We have seen something of how living communities develop through a series of successive steps, each preparing the way for the next. At each stage in this community development the existing vegetation may eliminate itself by building an environment more favorable to some other type. And with the changes in vegetation come changes in the animals that depend on it.

The world offers many different kinds of environment, and through the long history of life each one of these has changed from time to time, as conditions changed on the earth's surface.

So, through the ages, all forms of life are faced with a double challenge. They must not only excel in the struggle for existence, they must also adapt themselves to meet new conditions and still win out over their competitors.

In every form of life there are variations between individuals. These may be normal variations in inherited characteristics, such as length of hair or legs; or they may be in the form of a mutation, or new characteristic which other members of the family do not inherit as, perhaps, a new color of the eyes. Thus there is opportunity for new forms to develop, and perhaps to displace some older form in life's struggle. So the earth becomes a great testing laboratory where many forms of life try out their abilities against many others. Of all the different kinds of living things that have entered this race, many have fallen by the wayside and become extinct, while others have evolved to displace them with new and far greater powers of survival.

Fossils in the rocks tell us a fairly complete story about the evolution of life on earth from its beginnings. Even the length of time involved can be estimated in various ways, the most important being the changes in the radioactivity of the uranium buried in some rocks.

One of the most complete records of life's history lies exposed in the rocks of the Grand Canyon carved by the Colorado River in northern Arizona. The walls of this canyon stand a mile high, built of layers of rocks, each layer differing from its neighbors above and below, and each carrying signs that tell of its history.

The lowest layer, nearly 1,500 feet high, is estimated to be about two billion years old. Its structure tells the geologist that it came into position as molten rock. Vertical lines in it show that it was forced upward under great pressure, to form a mountain range. The top of this layer is a flat surface, cutting square across the vertical lines. This tells us that the old mountains were worn down by the forces of the weather until only this flat surface was left.

There is no visible record of life in the lowest rock. The records begin in the layers above it, which are made of sedimentary rock formed by the sediments of dust and sand, which are composed of particles worn from earlier rocks. The oldest layer, about 500,000,000 years old, is much younger than the lower rock. The sediments of which it is made were washed into the valleys and oceans, burying beneath that the remains of plants and animals as they lived and died through the ages. Pressure and chemical changes slowly cemented the sediments into rock again; and within these rocks are preserved fossil records of the oldest living things.

Layer after layer the sedimentary rocks were laid down, until today they lie piled three-quarters of a mile high above the old mountain base. In some places the rock structure shows that other layers were built and later worn away, but in the end result, in the towering wall that is left, each single inch of rock represents the work of about ten thousand years. The thickness of a single human hair spans perhaps ten years of growth. And this whole vast record of life's development covers only its early stages before the coming of the dinosaurs.

The lowest of these sedimentary rocks is made of limestone, mixed with sand and pebbles. Most limestone is formed under water, much of the lime coming from the remains of plants and animals which built it into their bodies from solutions in the water. It is very common to find the remains of shells and other aquatic life in such limestone rock. But the rock in our story carries the evidence of only one form of life. This evidence is shown by patterns of curving lines in the rock very similar to the outlines of colonies of microscopic single-celled algae, which grow today in many parts of the world. These tiny plants are among the smallest and simplest forms of life. Since there are no other records of life in this and other limestone of comparable age, it is a fair assumption that no higher life existed when it was formed.

Myriads of these microscopic plants filled the water with a tremendous supply of food. Some of them had the power of motion, and through the ages some cells developed the power to surround and absorb their neighbors, to make use of this stored food.

So there appeared the simplest animal life, single microscopic cells, like the amoeba, dependent on the plants to build food for them with the energy harnessed from the sun. Then, among the many living cells there developed some with differing functions. Some joined into colonies, and in these colonies some groups combined to carry out special functions, until higher forms of life gradually began to take shape. In the fierce competition for food those forms survived that developed the best means of catching food, or of protecting themselves from the appetites of others. Worms developed the ability to move on the ocean floor in search of food. Other creatures developed shells to protect themselves.

So, in the next layers of rock, we find the records of these new forms of life. Sand and mud and lime in the rocks tell the story of an old sea beach that slowly sank beneath the sea. In it are the records of the earliest sponges and shells and worms. The highest creatures at this time were the trilobites. Varying from less than an inch to a foot long, they had greater power to move and match their food than any other living thing. For many million years they were the dominant animals on earth until, at last, some new force in their changing environment overcame them, and they disappeared completely. As the story moves upward, in rocks about four hundred million years old, records appear of the earliest fish, clumsy and covered with armor plate but able to move in three dimensions and better equipped to catch their food than any other creature.

Above these, the rocks contain the sand and mud of an ancient flood plain, with the records of the earliest fern-like plants and the tracks of early land animals, not much more than fish with lungs and legs. These might have developed in the drying pools as the land rose from the sea, where some fish might have found the power to survive in the mud, to breathe air and move about on their fins, as some of the lung fish are able to do today.

The rocks above these contain strata of particles finely sorted and deposited by the changing currents of an old river. In this mud are the records of many ferns and cone-bearing plants. With them are the tracks of animals shaped like salamanders. And the wings of insects tell of nature's early experiments with flight, with its accompanying ability to follow and use a spreading food supply. With these records, too, are tiny pits left in the old mud by falling raindrops.

Higher still, the rocks are made of particles of sand rounded and smoothed by being blown in the wind. This sand is deposited in curving patterns that tell of wind-blown dunes, and here the tracks of many kinds of primitive reptiles and amphibians, able to live both on land and in the water, record the multiplying forms of life.

Above the highest rocks of the canyon there once were other layers that were later worn away by erosion, but to the east and north, parts of these layers remain, and they are exposed on the Painted Desert and in the walls of Bryce and Zion canyons. On the Painted Desert there are the tracks of the giant dinosaurs that ruled the earth as its largest and strongest creatures through millions of years. Then, apparently due to failure to adapt to some change in their environment, they too disappeared completely from the earth. This disappearance might have been due to the development of creatures that could find and destroy their eggs, or perhaps to the slowly cooling climate or some other change to which the cold-blooded dinosaurs could not adjust themselves.

Their place was taken by the mammals, warm-blooded creatures better able to adapt to changing climates—elephants, camels, ground sloths, and the saber-toothed tiger that fed on them and ruled the continent for long ages. Finally this tiger disappeared in its turn, perhaps because, with its huge and clumsy tusks, it was adapted to feed only on certain kinds of animals such as the ground sloth, and through this overspecialization failed to survive the loss of its needed food supply.

Along with the development of specialized and better adapted bodies there came the evolution of another quality, intelligence, that was destined to outweigh all other living forces in its effect on the world. Its beginnings were perhaps indistinguishable from the instinctive actions of hunting and eating.

For example, there appeared among the insects the power to perform feats of amazing skill. Henri Fabre, in his experiments with the Sphex wasp, proved the ability of this insect to perform the most delicate and exacting nerve operation on its grasshopper prey, paralyzing the legs, leaving the otherwise healthy grasshopper helplessly waiting to furnish living food to the young wasp when it hatched.

The wasp never used any grasshopper but one, the Ephippiger, and it always dragged its paralyzed victim to its nest in a special way by an antenna. If the antenna were cut off, the wasp appeared completely baffled, unable to conceive of the idea of dragging its meal by any other handle, such as a leg. Judging by this and many other experiments, Fabre concluded that although some insects possessed the most highly specialized instinctive skills, which they inherited without the need to learn from teaching or example, they completely lacked the ability to reason.

Other insects like the termites, bees, and ants, developed another ability, the power to organize group action, to assign different functions to different members of the group, and to enforce discipline with great efficiency. Interrelationships of this kind are carried to a very high point in the termite colony, whose members are divided into several different classes or castes. Some, having the power to reproduce, are responsible for the maintenance of the species, others are workers. Unable to reproduce themselves, the workers are charged with the building of the termite house, with its elaborate construction, containing chambers for special purposes, such as the rearing of the young and the storing of food, all with complex provision for ventilation and temperature control. The colony is protected by other members equipped as soldiers, with heads so adapted for their work that they are unable to feed themselves. Their lives are dependent on the attention of the workers who collect and prepare their food and then carry it to them. Some species of termites and ants go a step further than this and raise their own food. They build special chambers where, on prepared beds of chewed-up leaves, fertilized by their own excrement, they raise certain kinds of fungus, which becomes the entire food of the colony.

Some termites send out armies of workers to gather food for the colony. Blind, and helpless to defend themselves, these workers march in columns between files of soldier termites that herd them to their jobs and back to the colony with their loads of food. Any termite that outlives its usefulness is killed and eaten by its fellows, or starved to death by the simple process of denying it food.

One species of ant, Polyergus, has a class of workers whose only function is to raid neighboring colonies of other ants. Killing all those who oppose them, they will capture the larvae or young of the raided colony and bring them back to their own nest. Here they turn them over to a group of slave ants, who raise these young to become slaves in their turn. The slave ants will care for their masters who, unable to care for themselves, are entirely dependent for their food on the help of the slaves. In some tests it was found that one slave could keep about 10 of its masters well fed.

In this and many other kinds of organized action it might be possible for the insects to learn from the example of their elders, or perhaps to inherit the inborn knowledge as instinct, through a preformed set of nerve cells.

Some bees have developed the power to communicate knowledge of a less tangible kind. A worker bee, returning to the hive with a load of pollen, will perform a dance on the honeycomb. By different evolutions in this dance it can indicate to its fellow workers the distance and direction to the flowers where the food was gathered, and the workers can understand and follow these directions.

All these abilities indicate a power to recognize and to communicate images within a specific, limited range. Many animals have advanced beyond this.

I once watched through my field glasses as a family of young ground squirrels made their first experimental acquaintance with a wire cage-trap. It was a tubular, flat-bottomed affair made of coarse wire mesh, closed at one end, and with a hinged door at the other. In its center was a wire trigger. When a squirrel brushed against this, the trapdoor quietly closed. The trap was baited with a path of clover blossoms leading from the outside into its closed end.

Being far more succulent than the local diet of grass, these blossoms appeared very attractive to the young squirrels, who accepted the presence of the cage as an interesting and apparently not unwelcome addition to their neighborhood.

Within five minutes one of them had followed the

blossoms into the trap and the door closed behind him. I promptly transferred him to another cage and repeated my vigil. My approach to the trap was heralded by a sharp squeak of warning from one of the squirrels. When it had been set again, the youngsters were soon back to investigate, but they were now much more cautious about entering the trap. They tested the wire with their teeth and investigated every crevice in the mesh for an entrance other than the door. One of them crept in a few inches to gather the clover near the entrance and then retreated. Finally the appeal of the clover overcame caution and within fifteen minutes the door closed behind a second squirrel.

This was the last one to enter the trap. Those that remained apparently had no fear of the trap itself, running over it and trying to get in by every way except the door. Finally one of them dug his way underneath it and pulled the clover through the mesh.

These squirrels had exhibited many of the attributes of human intelligence—memory, the ability to understand cause and effect and to guide their actions on that understanding. They had used imagination to develop a new way to reach the clover and, knowing the danger, they had shown a willingness to take the physical risk to try it. And they had the power of communication. If the squirrels had the physical equipment to make sounds enough for a more complete system of talk, and to leave records for the accumulation of knowledge—if their lives were long enough to accumulate knowledge and experience comparable to man's—we would be in a better position to compare the actual quality of their intelligence with our own.

But one more ability has been developed, apparently possessed by man alone, the newest, most powerful of all the forces of life. That is the power to deal with abstract ideas, to analyze causes and effects, to recognize the principles that underlie them, to use these concepts as building blocks for new ideas in a process of creative imagination. From this imagination there has grown that restless urge which leads man to constant fresh activity—to heights of achievement when guided by wisdom, and to depths of stupidity when wisdom is lacking.

This is not just one more simple forward step in the process of evolution. It is a revolutionary change in direction. For the first time in the world's history it has given to a living creature the power to escape partially from the natural laws that control all other forms of life, and it has conferred the power to modify the environments on which all life depends.

In the development of a forest community we have seen the succession of different types of vegetation. Here each one in turn achieves a dominant position, only to modify its own environment so that it can no longer live there, and must give way to some other form. We have seen the deer of the Kaibab Forest multiply, through artificially produced prosperity, until they destroyed both their environment and themselves. In each case natural law stepped in to set a limit to the change of environment.

In his early days man was controlled by the same natural laws. When a community destroyed the land that supported it, the inhabitants had to move on to new land, or perish. Its powers of destruction were limited to its own environment. Today, with the development of transportation, a community can destroy its land and still live on it if there are means to bring food from outside. Thus, a community may destroy its land and continue to grow and destroy more distant land until it comes into competition with other communities which are also seeking new land to exploit.

In the development of an industrial civilization, human intelligence is confronting itself with a paradox. It has achieved almost unlimited power to destroy its environment, and almost unlimited power to multiply. These two together add up to almost unlimited compulsion toward destruction of environment, which eventually means self-destruction. While pursuing this course, the mass of mankind has become so far removed from the soil that supports it that it little realizes the problem involved. Orderly natural law has given way to unregulated chance.

There are probably few environments in the world today that are not in some way affected by the activities of man. Thus, human activity and the incentives that cause it are a vital part of most environments.

18. *human intelligence: evolution takes a new direction*

At a government experiment station in southern Arizona there is a field surrounded by a fence. Its surface is covered by a fine stand of grass, the good native grasses that the white man found here when he first drove his cattle in, growing knee high and thick, out of a deep, firm sod. This field has fed its share of cattle, but it has been protected from heavy overgrazing. Around this pinpoint of rich grassland there stretches a vast area of desert, the typical Arizona waste land that covers so much of the state today. Scattered over its hot sand are cactus and mesquite and burro weed and many of the great mounds of the kangaroo rat nests. Grass seed has been scattered experimentally on this land, too, all around the protected field. But it was found that on the bare soil, in only one year in ten did the earth retain enough moisture to give the seed a start. This whole desert area was once covered with grass like that in the field behind the fence.

I visited this grass-covered field one hot May morning and dug out a piece of sod from under the good grass. It came out a solid chunk of roots, six inches deep or more. Beneath it the soil was cool and black and moist, the kind to gladden a farmer's heart.

Outside the fence, five feet from this spot, I hacked out another bit of earth. It was dry and yellow, baked as hard as an adobe brick. Here there had once been moist black sod like that across the fence. But now the black humus and the protecting roots were gone.

Another time I visited this spot after a heavy summer thunderstorm.

Now I could visualize something of the process of destruction that had overtaken it. It was on a gentle slope. The storm had washed the higher parts of the desert clean of every scrap of debris. Here and there the tracery of fresh gullies had etched their outlines on the surface. Their courses converged toward an older, deeper gully whose vertical sides in spots were caving in. Here a mesquite, its roots undercut, had toppled in. A Cholla cactus had been washed against it.

Nearby, along the upper edge of the fenced field of grass, I could see the story of what happened. The grass stems were buried in fine fresh sediment two or three inches deep, washed off the slope of the nearby desert.

Against the outer edge of the grass stand was banked the lighter debris from the desert, little piles of dead mesquite leaves, bits of cactus, and the dried stems of dead burro weed.

Here was gathered all the new organic richness that had been accumulating on the soil. All this had been robbed from the desert and handed over by nature to grassland waiting to receive it and grow richer. If the grass had not been there to receive it, this rich potential raw material for meat and wool and leather would have been carried away with the water, to flow through the great gully, or *barranca*, as the Spanish call it, into the river below.

I once climbed down into this *barranca* with a man who had lived nearby for many years. We walked to a cottonwood tree growing on the bottom and looked up at its top branches, thirty feet above us.

They were on a level with the surrounding desert. My companion said to me, "Twenty-five years ago there were cattle grazing up there, where those cottonwood leaves are rustling. Those cattle and the cattle before them have murdered this land. In some places where twenty cows grew fat fifty years ago, one cow will starve today."

The story of this desert is rather typical for much of the range land of our Southwest. A good part of the water that supplied it came from the mountains, which were well forested on their upper slopes. The forests stored the rain and snow of the mountain storms, letting it out slowly to supply the land below. On the lower land, most of the rain comes in the occasional heavy thunderstorms of summer. To make use of these sudden downpours the land must have power to absorb water very quickly before it runs off the surface. The old stands of native grasses were well adapted to delay run-off and maintain a porous surface. In some places, grasses like the fescues grew in clumps, leaving room between for a great variety of nutritious herbs. We have already seen how the smaller animals worked among the forage to keep the surface of the land porous, and how the larger animals harvested the fresh grass tips but were kept in check by predators, so that they seldom cropped the grass close enough to damage the plants. Thus the range was kept healthy.

But when man began to raise on the range more cattle than it could support, a series of changes came over it, each change leading on to the next.

The first cattle to graze the range took the topmost crop of leaves, avoiding, as they always will, the less palatable stems and older leaves below. But these standing lower leaves were still grass and they could still feed cattle. To many cattle men they looked like unmined gold. And on the open ranges, if any man was wise enough to leave this golden factory unharvested, some other stockman driving his animals through was always ready to harvest it for him. Normally these perennial grasses ripen and stop growing in early summer. But with the overgrazing they must now send out new shoots in the later season, to maintain their life.

This late growth offers a great new food supply and lengthens the time when young grasshoppers can grow. In the old days these insects were controlled chiefly by birds, such as the Swainson's hawks which patrolled the grasshopper breeding grounds in their thousands, and by meadowlarks which hid their nests under the cover of the grass. But the hawks were killed off by the hunters, and with the shortening of the grass which had sheltered the nests, the land had lost the power to raise the meadowlarks which once protected it. So, given these favors, the grasshoppers multiplied, just as we saw the bark beetles multiply in the forest.

Under this treatment the grass roots, lacking the leaves to build their food, slowly grew weaker. The soil was compressed by the steady trampling of heavy hoofs faster than the soil animals could plow it. Losing its absorbent covering, it soon lost the power to absorb moisture from the rare rainstorms. Slowly it dried out, killing the weakened grass roots.

The perennial grasses were now replaced by the fast-growing annual grasses. Bearing heavy crops of seeds, they could establish new seedlings in any open spot that supplied the needed moisture. Having smaller roots, less able to protect the soil, they took the moisture and the space where the old grass might have recovered.

The heavy crop of seeds from the new grasses now offered an expanded food supply for rodents, and, like the grasshoppers, the rodents responded. Animals such as the kangaroo rat came and covered the ground with great mounds of bare earth, often surrounded by windrows of hulls from the grass seeds they had eaten.

Pressed by cattle, insects, rodents, and drying soil, the grasses in their turn gave way to other plants better adapted to withstand the pressure—plants with tough, dry, unpalatable leaves, such as the burro weed, plants that protected themselves by poison, such as the loco weed, or by sharp thorns like the cactus of many kinds. The drying earth could support only a scattered stand of these plants. In the open spaces between them the rain began to carry off the unprotected topsoil, and so the process of erosion accelerated step by step.

Water seeping through healthy soil or flowing gently over its surface will flow clear and sparkling. But if bare soil is exposed to the hammering blows of the raindrops, its surface is compacted and made waterproof. Finer particles of soil are spattered into the air and, falling, are carried along by the flowing water. With the waterproofing of the soil's surface, the water that once seeped into the ground now joins the surface flow, and with the added flow the ability of the water to carry soil particles increases. If its speed is doubled it can carry particles not just twice as big but sixty-four times as big. It begins to cut small gullies. Here the water collects in larger streams, and again its speed increases. With each doubling of its velocity, the size of the rocks it can carry is again multiplied sixty-four times, until the dust particles that it started with are joined by huge boulders that act like battering rams against the banks as they are hurled along. So the banks are torn away. In some places the old grass-covered soil has gone completely, giving way to bare rock. There, the soil reservoir has not just been damaged—it is destroyed, never to return in the lifetime of men now living.

With the destruction of the soil, the flow of the streams became more irregular and violent—floods followed the rains; then, with the water reserves gone, came scorching droughts. The supply of water to the valley became less dependable, and with this change the life of the valley began to deteriorate. The process of weakening the vegetation spread its disastrous efforts to produce the desert.

This man-made desert, surrounding the fenced field of rich grass which we saw earlier, gives us a picture of something entirely new in the history of life on this earth—new with the coming of man. This new picture shows us the impact of misused intelligence on environment. It is not just the impact of intelligence in southern Arizona. It is a rather typical phenomenon for much of the North American continent—for much of China, India, Asia Minor, southern Europe, parts of Africa. The small, protected, healthy field in the midst of this desert is perhaps a picture of the next step to come in human evolution, the development within man of understanding to govern his intelligence with wisdom. On this small plot a single strand of wire has fenced out the grasshoppers and cactus and the marching desert, because it has fenced out the surplus cattle that prepared the way for them. It was put there by a man with understanding.

When we think of this desert it is easy to blame human greed; but it is well to remember that this answer is too simple. Greed no doubt played its part. So did imagination, initiative, hard work, the willingness to risk life and savings to gain independence. These are some of the things that we regard as among the higher attributes of man. They were vitiated by ignorance, by lack of established responsibility, and by the pressure of competition which led to the crowding of the range beyond its carrying capacity, until today in many places that carrying capacity has practically gone.

This process of land destruction may go on wherever the natural covering of the soil is broken without proper provision for its protection. Huge areas of the earth have been totally destroyed in this way since man first learned to harvest crops.

It has been said that scientists have learned more about proper land management in the first half of the twentieth century than in all the earlier years of human existence. Underlying this new knowledge there is one basic fact—that the surface of the earth is a highly complex living organism. It can be harnessed and made very productive if it is properly treated, or, like the goose in the fable, it can be opened up and all the golden eggs removed.

Through most of his history, man has been busily digging out the golden eggs. To change his old inherited ideas about land use while there is still time becomes a colossal task. Fortunately, great forward strides are being taken. We have learned that the ability of the soil to absorb moisture can be maintained by cultivating along the level contours of the slopes so as to form ridges and troughs to hold water. We have learned that the quality of the soil can be preserved by harrowing or disking in mulches to renew the humus, to keep the surface porous and to protect it from the wind. We have learned that cover crops can be planted to hold soil in place against erosion when it is not being cultivated. We have learned that fertilizers can help to maintain its productivity. We have learned that forests can be managed so as to insure the preservation of cover on land after cutting, and to provide for a new crop of trees to replace the old.

But with all our new knowledge of land management there is still an enormous amount to be learned, and the greatest task of all is to provide the user of the land with incentives to proper management. For, every user of productive land is in fact a trustee, responsible for the lives and the future of the human race, which must still depend on that land for support.

19. *man in the web*

Through the city of Los Angeles runs a canal lined with concrete. During some parts of the year this canal carries a thin trickle of water, the last vestige of what was once the Los Angeles River. Today the entire normal flow of this river runs through the pipes of the city water system to do its share in meeting the city's needs. Long ago the city outgrew this supply of water, but this was not too serious. There was a huge supply underground. Shallow wells produced enough for every additional need. As seemed right and proper, every man owned all the water beneath his land, and he was free to pump as much as he desired. The thriving city and its surrounding communities grew. Households and factories and irrigated farms all took their growing share of water. Slowly the level of the underground water began to sink. Wells were dug deeper, and still the water dropped. Like the river on the surface, this underground supply was fed by the flow from nearby mountains. It was so vast in extent that it gave the appearance of being inexhaustible. But when the demands outgrew the streams that fed it, this river too could go dry like the surface river, and it became apparent that the wells were drawing from the ground more than its entire income of water.

New sources must be found, or the community must shrink back to the size that the supply would support, and at fearful cost to those who must go.

A great new aqueduct was built into the mountains to tap the flow from the distant Owens River, and still the city grew. At last the project was conceived of bringing in water from the Colorado River. This meant carrying it through an aqueduct over more than 220 miles of desert—a perfectly practicable engineering feat. But six other states and, also, our southern neighbor, Mexico, all depended on the Colorado. All had rights to the use of the water. All claimed needs for it which, put together, totaled far more than the entire supply. The cost of the engineering works to make it available was beyond the means of any one of the states.

The work could only be done by the states acting together, with the cooperation of the federal government. This would require money drawn from the taxpayers of the whole nation, and it could only be done by reconciling the claims of the interested states.

A compromise was finally reached and work begun on plans for a series of dams to store the river's flow in the flood seasons and make it available through the dry seasons. This would give a supply that could be depended on when it was needed throughout the year.

The state of Colorado even went so far as to tunnel through thirteen miles of solid rock under the mountains of the continental divide, to divert water from this river across to the eastern side of the state. There it joins the Big Thompson River to flow down hundreds of miles of waterways in the great river system that feeds the Missouri. Thus, water that would have reached the Gulf of California was made to flow into the Gulf of Mexico. And on this journey its value can be multiplied by re-use as each drop of water gives its service again and again to the homes and towns and cities along its course. Each user in turn takes it from the river, uses it, and defiles it, then returns it with its load of refuse to the river for the next user to get from it what service he can. We must remember that one of water's chief uses to man is as a solvent, cleaner, and conveyor of wastes.

As the early cities grew along the river courses, this pollution was not a serious problem, for the wastes from each city were diluted by the flowing water, oxidized by the bacteria, used as fertilizer by the water plants, and filtered through the river sands and gravel, so as to reach the next user in fairly clean condition. But with the multiplication of cities and their discharges, the water became filled with an unsupportable load of poisons from the factories, offal from the slaughterhouses, raw sewage from the homes. These killed the cleansing plants, used up the purifying oxygen in the water, and clogged the filtering gravels with filth.

And so, today, the water supply for many of our cities enters the city water system as a dark chocolate-colored fluid, straight out of the sewers and factories of its neighbors upstream. In one midwestern city some tests of the water showed that, during the period of low water in the winter, it was one-half straight sewage. Later in the season, when the river filled with run-off from the spring rains, it was a thin, liquid mud, made from good topsoil washed off the improperly cultivated farm lands above.

By thorough and expensive treatments, the mud and the visible sewage can be removed from the water and its remaining load of bacteria killed by disinfectants before it is distributed to citizens for drinking, washing, and cooking. But then it is discharged back into the river, boiling out from the city sewer as foul and offensive as when it entered. The sewers of the city have become an integral part of the watershed that supplies the cities downstream.

This pollution of the rivers is as truly a destruction of a basic natural resource as is the overcutting of a forest or the wrong management of good land. Like the land and the forest, the water can be restored to usefulness at great expense if it is not too far gone. But, as in their case, this destruction is unnecessary. It should be cleaned by its users before its return to the river. This can be done by proper treatment. The wastes and chemicals thus recovered are valuable for fertilizer and other uses. But it is an expensive process, beyond the immediate means of many communities and factories, so the nationwide drive toward improvement is slow and painful.

Another destructive effect of this fouling of the streams lies in the filling of the nation's reservoirs with silt off eroding hillsides. The erosion is chiefly the result of improper management of the watersheds. We have seen that it is adding about 700,000 tons of silt each day to Lake Mead above Hoover Dam. This siltation is not merely destroying a huge reservoir—it is working like a hidden cancer to destroy the water supply on which the life of many great communities is being built. What will happen to them when it is gone? In the case of Hoover Dam it may be possible to build new dams on new locations along the river, at the cost of billions of taxpayers' dollars. In the case of many other dams this will not be possible when the dam-created lakes are filled with silt, for there are no other available sites for new dams. They form a basic natural resource. Once destroyed, they are gone forever.

There is much still unknown about the proper management of watersheds, and since the conditions on each will vary with soil and slope, climate and vegetation, the management required on them may differ widely. But in every case the first requirement is to protect the soil, which is nature's reservoir for storing water where it falls.

So, we see the river and the lands of its watershed

as a great living organism, with its heart in the mountains that supply its life blood. This blood flows out through the streams that form the arteries above ground and below, coming down from a hundred thousand hidden sources—the mountain springs and meadows, the patches of moist woodland with the porous soil beneath them, the shaded snow banks and the afternoon thunderstorms, the flow of every raindrop held back by the delaying stems of grass and flowers, absorbed by bits of rotting wood, filtering into the soil through a million root tunnels and worm holes, delayed, but slowly moving down the hillside through the soil, to bring a steady, even flow of life to the great functioning body of civilization in the valley.

Every action that affects the lands of the watershed has its direct influence on the functioning of the whole organism. The growing leaf that shades the snow to delay its melting is doing its microscopic share to give an even flow of water through the summer. Combining its influence with a hundred billion other leaves, it may determine the success or failure of the harvest in the valley.

We may see and appreciate the devastation caused on the watershed by a carelessly dropped cigarette. But what of the sheep whose hoof compresses a worm hole or destroys the protecting grass roots on the hillside? Nature may repair these tiny bits of damage if they are kept within limits with which she can cope. But when too many sheep combine to lay a hillside bare, they set the stage for rivers filled with silt, and for next season's floods and droughts over huge areas of distant lands.

The people who use the watershed hold in their hands the lives and well-being of the millions who depend on it. Every action that affects the health of the watershed becomes a matter of vital concern to those millions, and to all the communities of plants and animals that together make up the whole.

And so we see that every community may be divided into four parts: first, its members with their immediate environment; second, the distant and unknown lands that send out their influence by stream and wind, by wing and padded foot to affect the local environment; third, the actions of men whose influence spreads out to affect in some way nearly every community living on the earth; and last, but most important, those influences that mold the minds of men, giving them the incentives to wise or unwise actions; for in the end these lie at the very center of earth's great web of life. To support this web, the soil must be maintained alive and functioning.

20. *summary*

As we draw to the end of this story of the web of life, it will be helpful to stop for a moment and consider its meaning. A glance at the picture we have seen brings out five points that seem especially significant.

The *first* of these lies in the endless ramifications of the environment that surrounds every living thing. To see this environment as a whole, we find that our story must cover a wide sweep. So we started out with a simple account of the raw materials which go into the making of an environment. We ended with a look at the forces which mold human incentives and behavior. Those two cannot be separated. With either one of them missing, the picture is distorted and incomplete.

The *second* important point to be emphasized is the teamwork and delicate balance existing between the different forces that keep life functioning on this planet. This was shown very simply in the story of the interdependence of plants and animals in Chapter 4. We saw how plants take nature's raw materials of earth, water, air, and sunlight and organize them into living substance. But without the help of animals, the plants would long ago have organized themselves out of existence, for they would have locked up in the earth the world's available supply of carbon and then died of starvation. Fortunately, the animals came along in time to eat the plants and thus release the carbon and keep the cycle functioning. The carbon not thus released is still available to us today in the form of coal. Plants therefore build food that makes life possible for animals, and in turn depend on the animals to keep the cycle of life moving.

The *third* point concerns the constant progression of life. The simplest plants and animals build the soil and prepare it for storing water and mineral solutions. This makes life possible for higher forms of plants and for animals whose lives are controlled by the vegetation that gives them food and shelter. The plant communities grow, modify their environment, and at last can no longer survive in the surroundings that they themselves have built. They must give way to other species, better adapted to the new conditions. And, of course, the animals too must change with the plants. This succession of new forms continues in orderly progress until a stable condition of balance, or climax, is reached. The dominant trees or other plants that form this climax can hold their place against all competitors. But we observe that they could never have got a foothold until the lowly pioneers had prepared the way for them.

We have seen that the dominant climax plants will vary with each different type of soil and climate. And their adaptation to any given area is so delicate that a seemingly very insignificant thing may wipe out the entire community. A local windstorm weakens the natural defenses of a group of spruce trees and neutralizes the defenses of their protecting birds. This transforms the local bark beetle from a petty nuisance into a raging epidemic, capable of wiping out the forest resources of a great state. With the forests go their specialized animals, and the whole community of life in the area is transformed.

This is a quick and drastic change, easy to see. But there is another type of long-term evolution that we cannot see so easily. New races of animals and plants have developed and dominated the earth for millions of years, only to disappear as their environments have changed, or new and better adapted forms have come in to drive them out.

This brings us to the *fourth* point in our review. With the development of human intellect, the evolution of life on this planet appears to have reached a turning point. In the past, all forms of life were subject to automatic natural controls. No species could succeed or multiply beyond a certain point without encountering controls that would eventually put it back in its proper place or even eliminate it, and the great stream of the world's life would go on undisturbed.

Today this condition has been changed. Through the use of his intellect, man has partially escaped from the controls of nature. He has achieved almost unlimited power to multiply his numbers and at the same time to destroy the world's resources that might have supported him. Under the domination of his intellect, the world's life, and the environment of that life, seem to have reached a crossroads, and the choice of direction is for the future to decide.

This brings us to the *fifth* of our main points—the basic ingredient of our story. Scientists have proved that it is possible to use the life-supporting natural resources of the world without destroying them. These resources can be renewed and made more productive with use. This is a triumph of the human intellect. The great and deciding test, however, still remains—whether man can coordinate knowledge into understanding and build within his heart the incentives and the wisdom to use these new-found powers wisely, and with responsibility, for the common good.

The thought that I would like to leave with the

reader is beautifully expressed by Alan Paton in his book *Cry, the Beloved Country:*

"The grass is rich and matted, you cannot see the soil. It holds the rain and the mist, and they seep into the ground, feeding the streams in every kloof. It is well tended, and not too many cattle feed upon it; not too many fires burn it, laying bare the soil. Stand unshod upon it, for the ground is holy, being even as it came from the Creator. Keep it, guard it, care for it, for it keeps men, guards men, cares for men. Destroy it and man is destroyed."

bibliography

Allee, W. C., and others, *Principles of Animal Ecology*. Philadelphia, W. B. Saunders Co., 1951.

Balfour, E. B., *The Living Soil*. New York, The Devin-Adair Co., 1950.

Benson, A. L., *The Story of Geology*. New York, Cosmopolitan Book Corp., 1927.

Cocannouer, J. A., *Weeds: Guardians of the Soil*. New York, The Devin-Adair Co., 1950.

Dunbar, C. O., *Historical Geology*. New York, J. Wiley & Sons, 1949.

Elton, Charles, *Animal Ecology*. New York, Macmillan, 1927.

Howard, Sir Albert, *The Soil and Health*. New York, The Devin-Adair Co., 1947.

Kittredge, Joseph, *Forest Influences*. New York, McGraw-Hill, 1948.

Langlois, T. H., "North American Attempts at Fish Management." *Bulletin of Bingham Oceanographic Coll.*, Vol. II, art. 4, May, 1948.

Leopold, Aldo, *Game Management*, New York, Scribner's, 1933.

Lyon, T. L., and Buckman, H. D., *The Nature and Properties of Soils*. Toronto, The Macmillan Co., 1943.

Murie, Adolph, *Ecology of the Coyote in the Yellowstone*. U. S. Govt. Printing Office, 1940.

Netboy, Anthony, and Frank, Bernard, *Water, Land and People*. New York, Alfred Knopf, 1950.

Odum, E. P. *Fundamentals of Ecology*. Philadelphia, W. B. Saunders Co., 1953.

Oosting H. J., *Study of Plant Communities*. San Francisco, W. H. Freeman & Co., 1948.

Osborn, Fairfield, *Our Plundered Planet*. Boston, Little, Brown & Co., 1948.

Platt, R. H., *Our Flowering World*. New York, Dodd, Mead & Co., 1947.

Platt, R. H., *This Green World*. New York, Dodd, Mead & Co., 1942.

Sears, Paul B., *Deserts on the March*. Norman, Okla., Univ. of Oklahoma Press, 1935.

————, *This Is Our World*. Univ. of Oklahoma Press, 1937.

Smith, G. E., and Albrecht, W. A., "Biographical Assays of Soil Treatments." *Soil Service Society of America Proc.*, 1942, Vol. 7, p. 235.

Tannehill, I. R., *Drought: Its Causes and Effects*. Princeton, Princeton University Press, 1947.

Teale, E. W., *The Insect World of J. Henri Fabre*. New York, Dodd, Mead & Co., 1949.

Transeau, E. N., and others, *Textbook of Botany*. New York, Harper & Brothers, 1940.

Vogt, William, *Road to Survival*. New York, William Sloane, 1948.

Waksman, S. A., *Principles of Soil Microbiology*. New York, J. Wiley & Sons, 1952.

Wickenden, Leonard, *Make Friends With Your Land*. New York, The Devin-Adair Co., 1947.

Williams, Dr. Arthur B., *The Composition and Dynamics of a Beech-Maple Climax Community*. Cleveland Museum of Natural History.

Woodbury, Angus M., *Principles of General Ecology*. Toronto, Blakiston Co., 1953.

"Local Climate in the Copper Basin." *U. S. Dept. of Agriculture, Circular* 774.

"Mean Precipitable Water in the United States." *U. S. Weather Bureau Tech. Paper #10.*

"Silt Flow in the Colorado River and Its Relation to Irrigation." *U. S. Dept. of Agriculture, Tech. Bulletin #67*, Feb., 1928.